THE INDOOR RADON PROBLEM

THE INDOOR RADON PROBLEM

Douglas G. Brookins

Columbia University Press
New York

LIBRARY OF CONGRESS CATALOGING-IN-PUBLICATION DATA
Brookins, Douglas G.
  The indoor randon problem / Douglas G. Brookins.
    p. cm.
  Includes bibliographical references.
  ISBN 0-231-06748-8 (alk. paper)
  1. Radon—Environmental aspects. 2. Indoor air pollution.
3. Radon—Environmental aspects—United States. I. Title.
TD885.5.R33B76 1990
628.5'35—dc20   89-22171
                                              CIP

COLUMBIA UNIVERSITY PRESS
New York Oxford
Copyright © 1990 Columbia University Press

Printed in the United States of America
10 9 8 7 6 5 4 3 2 1

Book design by Ken Venezio

For Judith

Contents

Preface

The importance of indoor radon, here taken to include radon-222 and its progeny, as a major health concern is relatively new. The relationship of high radon levels and a high number of lung cancers in uranium and other miners has been known for decades. Similarly, the effect of other alpha emitting isotopes on human tissues, due to medical and accidental occupational exposures resulting in lung cancers and other diseases, is also well documented.

Only recently, however, have people become aware of the more subtle health risk of radon that accumulates in dwellings under normal living conditions. Yet this radon, which emanates from soil and bedrock, from building materials, and even from water and air, is a major cause of lung cancer fatalities. In the United States alone some 20,000 lung cancer fatalities due to indoor radon occur each year. The figure may actually be higher, but cases of combined smoking- and indoor radon-caused lung cancer simply wind up as deaths related to smoking.

Yet the high number of indoor radon-induced lung cancers need not occur. If a dwelling is found to contain elevated radon levels, it can be subjected to remedial methods ranging from the very simple to the very complex—and usually with good results. The problem, however, is that most people in most countries are unaware of the magnitude of the problem, and the means to determine areas of high radon potential are not widely available. Despite this, there are ways to address this important issue.

The soil and rock on which we live produce most of the indoor radon. There is a complex interplay of many variables that result in radon loss from soils and rocks and thus entry into dwellings. But all facets of the problem are approachable, if not resolvable. It is now rewarding to see the research efforts on indoor radon.

This book is not intended as an in-depth research work. There are other books that meet those criteria, many of which will be included in the list of additional readings and references at the conclusion of each chapter. This book is intended to reach the lay public, and it assumes that the reader has some knowledge or access to high school chemistry. In case the reader's education has not included such material, a glossary of terms is also provided.

My intent here is to provide the public with a better understanding of radon from its origin by radioactive decay from uranium release mechanisms from soils and rocks, uranium distribution and geochemistry in the earth's crust, radon entry into dwellings, radon detection methods, and radon mitigation processes. Two chapters address radon surveys and special problems in as well as outside the United States. One chapter deals with application and use of radon emanations for prospecting and other studies, and another chapter is about risk of indoor radon as compared to other sources of risk. Finally, a short chapter states my conclusions and recommendations.

I have benefited from discussing aspects of this book with many individuals and include the kind cooperation of several who provided me with reprints, reports, and other technical radon information. Special thanks are due to Dr. A. V. Nero, Jr., of Lawrence Berkeley Laboratory, to Dr. Richard Toohey of Argonne National Laboratory, and especially to Dr. Douglas Mose of George Mason University.

The patience and camaraderie of Mr. Edward Lugenbeel, executive editor of Columbia University Press, are gratefully acknowledged, along with deep thanks to Ms. Judith Binder, who edited the numberless drafts, and provided encouragement during the period in which this book was written, and to Mr. Aaron Johnson who drafted some of the figures.

Finally, I wish to thank my students and colleagues in Albuquerque for their assistance on radon work, their input on the subject, and their support. I wish to especially recognize Dr. William Mansker, Mr. Robert Beard, Mr. Ed Calvin, and Mr. Yehouda Enzel for their contributions. For providing financial support in order to conduct research on indoor radon, I thank Dr. Theodore Kauss, for the Frost Foundation, and the University of New Mexico Research Allocations Committee.

THE INDOOR RADON PROBLEM

Introduction

The "Indoor Radon Problem"—what is it? Apparently it means different things to different people. To some there is no problem. In Montana
and in Germany, Yugoslavia, and elsewhere in Europe, high radon-
emitting areas are thought to have therapeutic qualities for the cure of
arthritis, headaches, influenza, sores, tumors, and many other maladies,
although none of these claims for cures have ever been verified by
scientific study. It is conceivable that the excess gamma radiation in
these radon spas may have some beneficial effects, as natural chemother-
apy so to speak.

In the early part of the twentieth century, a toothpaste called Radi-
ogen was marketed. The toothpaste contained radium, the immediate
parent to radon. It was supposed to stop plaque from forming. Radium-
bearing laxatives were also marketed about the same time—good for all
kinds of maladies, including rheumatism, neuritis, and so on. The United
States government banned radium and radon containing products in
1938, although at least one firm tried to market radium bearing contra-
ceptive jelly in the early 1950s.

For many people, however, indoor radon poses a very real health
threat, mainly in the form of lung cancer; but this threat is difficult to
"sell" to the public. On the one hand, radon is colorless and tasteless,
and its presence can only be detected by chemical or radiation-measuring
techniques. On the other hand, cigarettes, the major cause of lung can-

cers worldwide, are more easily associated with ill health since one can see the foul fumes, smell the acrid smoke, notice the nicotine stains on fingers, and so on; and, by law, warnings about the negative health effects of smoking appear on all cigarette packages and advertisements. Not so for radon. Radon comes from the earth. Its release to the environment is a natural phenomenon that may be enhanced or retarded by people's activities. The only warnings that one can make are by educating the public about radon, and that is one of the primary purposes of this book.

Radon is an inert gas like argon, neon, krypton, and xenon, and under normal conditions it does not react chemically with other substances. Radon is also radioactive, and it decays by emitting radioactive particles to create the radioactive daughter isotopes of polonium, bismuth, and lead, which are known as radon daughters. These radon daughter isotopes are the actual culprits in lung cancer, although, since they are formed from radon, they are included under the heading "indoor radon problem." While the mechanisms of decay, and the elaborate decay schemes for radon isotopes and radon daughters are covered in detail in chapter 3, here I will briefly introduce the subject.

For this book, I consider most indoor radon to be radon-222, i.e., that isotope of radon that contains 86 protons and 136 neutrons in its atomic nucleus (86 + 136 = 222). Two other isotopes of radon are radon-220 (so-called thoron) and radon-219 (so-called actinon). The ultimate parents of these radon isotopes are isotopes of radioactive uranium and thorium. Uranium in nature consists of the very abundant uranium-238 (99.3 percent of total uranium) and the much less abundant uranium-235 (0.7 percent of total uranium). Of the radon isotopes, radon-222 and radon-219, the radon-222 forms from uranium-238 and is therefore much more abundant than radon-219, which forms from uranium-235. Radon-222 (thoron) forms from parent radioactive thorium-232, which, while more abundant than uranium, is not as mobile and not as much of a threat as radon-222, which is formed from uranium-238.

Most uranium and thorium in the earth decays by a series of steps until nonradioactive, or stable, isotopes of lead are formed. Uranium-238 decays eventually to lead-206, uranium-235 to lead-207, and thorium-232 to lead-208. All of the other isotopes formed from parent uranium and thorium isotopes are radioactive and are called intermediate daughter products. Thus, any of these can be both a radioactive parent and a radioactive daughter. In the uranium-238 chain, for example, radium-226 is the immediate radioactive parent to radon-222, so the latter is a daughter to radium-226. Yet radon-222 decays to polonium-218, so radon-222 is also a radioactive parent. Most of the intermediate radioactive daughter isotopes formed by the decay of uranium-238, uranium-235, and thorium-232 do not escape from the mineral in which they are formed. In fact, geochronologists can date rocks and minerals by using the so-called uranium-lead and thorium-lead methods, one measures the amount of the parent uranium or thorium in the sample, the amount of stable lead isotopes that have accumulated, and, since the rates of decay of the uranium and thorium are known, the age can be calculated, assuming no loss of parent or any daughter. However, as will be pointed out in chapter 2, small amounts of loss of some of the intermediate radioactive daughters does occur from time to time. In the uranium-238 decay chain (see chapter 3), loss can occur from thorium-230, uranium-234, radium-226, and especially radon-222. Because it is a gas and inert, radon-222 is not tightly bonded in the mineral in which it is found. Therefore, by weathering, heating, oxidation, and other means, some radon is lost from rocks and minerals. Much of this released radon-222 makes its way to the surface of the earth and is then released into the atmosphere. The earth's radon budget is discussed in chapter 2 in detail.

At the earth's surface atmosphere, the risk from radon or radon daughters is very slight. If, however, one builds a home or other dwelling, then the release of the radon to the atmosphere is blocked, and, along with other gases, the radon accumulates in the dwelling. Even in well-ventilated dwellings, the indoor level is greater than that encountered out of doors. For homes that are especially well constructed and insulated, radon may readily accumulate to high levels.

As discussed in chapter 3, radon is radioactive and decays with a half-life of 3.8 days. That means that one-half of the original radon present in any particular volume will disappear by this decay in 3.8 days and one-half of that remaining will decay in the next 3.8 days, and so on. But in a volume such as that of our lungs, we continually replenish the radon supply from the air as we breathe. We inhale a certain amount of radon, and we exhale most of it. Yet a very small dose does undergo radioactive decay in our lungs. The real problem, though, is from the radon daughters that we inhale along with the radon. These radon daughters are present everywhere in the air around us just as radon is. So, if the air has a high radon gas level, it will also have a high radon daughter level. Thus, when we inhale large amounts of radon, we also inhale large amounts of radon daughters. These latter isotopes are those responsible for damage to lung tissue. The radon-222 we inhale that undergoes radioactive decay in our lungs is a minor contributor because this isotope is only weakly radioactive, i.e., it imparts very little energy or effect on the surrounding tissue. However, two isotopes of polonium, polonium-218 and polonium-214, formed from radon-222 are more highly radioactive and their decay has a pronounced effect on tissue and cells. This is covered in detail in chapter 4.

There is a popular misconception that one ought to be concerned about radon only if living near uranium ores, uranium mills, uranium mill tailings, and other areas where uranium is present in abundance. Not so. Uranium is a trace element in the earth, but it is remarkably widespread. All rocks contain some uranium, and many have more uranium than others. Just the gradual loss of small amounts of radon from host uranium minerals over long periods of time can cause radon to be lost in quantity from favorable areas. How this loss occurs is covered in chapter 5. Briefly, however, the radon may be dissolved in water and transported to some release point to the earth's surface, or it may move in its gaseous state along grain boundaries, cracks, etc., until the atmosphere is reached. Regardless of the mechanisms involved, large amounts of radon do make it to the air we breathe.

It has been known for over thirty years that some uranium miners suffer above-average incidences of lung cancer, and radon, along with radon daughters, is the cause in some of these cases. This picture is not clear because many miners smoke, and the risk from smoking is so high that it in turn masks the risk from the high-uranium area radon. Yet the public is pretty much aware of the risks that uranium miners have suffered, especially during the 1950s, when mine ventilation and monitoring for radon were poorly carried out. Now, with use of respiration masks, excellent mine ventilation, and constant monitoring, the incidence of radon-induced lung cancer among uranium miners is much lower.

What the public has not known, however, is that many groups of miners suffer radon-induced lung cancer because this gas can accumulate virtually anywhere underground. In some areas, such as the fluorspar mines in Newfoundland, tin mines in the United Kingdom, niobium mines in Fennoscandia, tin mines in China, and many other types of mines in other parts of the world, poor ventilation and lack of monitoring have led to high incidence of lung cancer among miners. This is not surprising since routine radon examinations are rarely carried out in other types of mining. Thus, high radon levels can go undetected for many years. Agricola in the sixteenth century mentioned that disease of the lungs affected many miners in Europe. Now sophisticated studies of uranium miners in the United States, Czechoslovakia, China, and elsewhere, coupled with the fine studies of other types of miners, allow researchers to determine good dose to adverse health effects relationships. This is covered in detail in chapter 4, where it is noted that the data are all for high radon doses. Risk relationships are very complex, and in chapters 1 and 4 I attempt to extrapolate the high-dose risk data to lower doses.

Regardless, the concern over indoor radon as a problem has been somewhat parochial in the United States. Some concern was voiced in the Grand Junction, Colorado, area where uranium mill tailings had been used to make construction materials (mainly bricks) for homes in that city. Unfortunately, even though the attempt to use the waste tailings was considered practical at that time, the tailings had not been completely stripped of their uranium during the milling process. Most milling methods are 90 to 97 percent efficient, and, while that sounds pretty good, it must be remembered that uranium ores, by definition, are highly enriched in uranium relative to nonore rocks. Thus a typical uranium ore may contain 0.5 weight percent uranium, which is 5000 parts per million (ppm), the units in which uranium is most often discussed. If milling removes 97 percent of this uranium, the waste rock still contains 150 ppm uranium after the process. This 150 ppm is much greater than the few ppm usually found in rocks (i.e., most rocks range from 2 to 10 ppm), and in turn, the waste rock is potentially a greater source of radon release. When the resultant high levels of radon and radioactivity in the Grand Junction homes were discovered, all the suspected dwellings were monitored, and, eventually, many were razed and new homes were constructed. Yet the Grand Junction experience did not catch the fancy of the press or public in terms of potential risk of other sources of radon.

Radiation units will be discussed in chapter 1, but I must introduce the unit that most commonly represents radon. The curie, for which the symbol is Ci, is broken down to the more convenient term picocurie or one trillionth of a curie and represents the tiny amounts of radioactivity given off by radon. From a survey of the homes in Grand Junction in which no mill tailings had been used, the U.S. government decided that 4 pCi/L (four picocuries per liter of air) was a reasonable level to achieve. However, even the risk from 4 pCi/L is not negligible (see chapters 1 and 4).

The Three Mile Island nuclear power plant accident in 1979 managed to focus a great deal of attention on radiation at all levels; I especially remember one congressional committee hearing where the amount of released radiation was in dispute. It was then pointed out that the background radiation in the room of the hearing was higher than that of the Three Mile Island release in question. Monitoring houses and other buildings was being carried out at about this time, and on occasion, houses with elevated indoor radon levels were encountered, yet there was no hue and cry about indoor radon as a problem.

The indoor radon problem was perhaps masked by a greater concern over other gases that might accumulate in homes, such as nitrogen oxides, formaldehyde, carbon monoxide, and others. Many of these are due to people's contributions, either from their building materials or from the contents of their homes. But this is not the case for radon.

The scientific community has been aware of indoor radon as a potential health problem for several decades, but its findings were not readily communicated to the medical profession nor to the lay public. In such nations as Canada and Sweden, research on the indoor radon problem has been ongoing since the mid-1970s, earlier in some cases. Yet in the United States, despite major contributions by a large number of workers, most of whom will be referenced in this book, the public has remained uninformed.

In 1980, the Swedish government published its findings on indoor radon and noted that many new buildings showed elevated readings. This was found to be due in part to use of a high-uranium radium shale (alum shale) in making cement, in part to excess insulation (to keep energy bills down), and in part to somewhat uranium enriched granitic rocks in the surface and subsurface. A short time later, in the United States, the Watras case came to light.

Stanley Watras, an employee of the Limerick Nuclear Power plant in Pennsylvania who lives in nearby Boyertown, had been setting off radiation monitoring devices at the power plant. Decontaminating him sometimes took more time than he worked. Finally, he requested that he be monitored just before his work shift. Again, he set off the alarms. Thus, he knew he was bringing the radioactive material with him to work and not obtaining it from the job at the power plant. Study of his home revealed indoor radon levels of over 2000 pCi/L (equivalent in lung risk to smoking 135 packs of cigarettes per day!)—the highest reading ever obtained for a private dwelling in the United States. Watras and his family were advised to move from their home in January 1985, even leaving behind new Christmas presents because of the extreme contamination. The Pennsylvania Department of Environmental Resources then studied the whole area and found that numerous homes built over the Reading Prong—a fairly uraniferous Precambrian rock body—contained high radon levels. Yet nearby, on rocks over much younger sedimentary beds, no such levels were noted. The Watras home was subjected to a concentrated clean-up effort, and the radon levels were successfully lowered to acceptable limits. This story has an apparently happy ending, except, understandably, the Watras family is concerned about their years of residence in their home before the radon was detected. The question remains: How about the numerous homes with high radon levels that are unknown to the people who live in them? This issue, with respect to the United States, is addressed in chapter 7.

Now, finally, the indoor radon problem is getting the attention it deserves from federal, state, and other sources, yet it is more timely and topical a subject in certain parts of the country than in others. In Pennsylvania, New Jersey, New York and parts of Washington, Colorado, and South Dakota, indoor radon is appreciated as an important potential problem for public health; in much of the rest of the country, unfortunately, this is not the case.

Curious about the publicity surrounding the Watras case, Roberta Baskin, a television reporter in Washington, D.C., decided to gather information about radon in Washington and the surrounding area. After approaching the U.S. Department of Energy (DOE), which felt the testing would be "uneventful," she, with DOE's help, placed charcoal detectors in fifty houses with basements. In the fall, detectors were placed in basements and in living areas. In the winter months, the test was repeated. All the houses selected were from the District of Columbia and nearby Virginia and Maryland. To her surprise, over half the houses tested yielded results above 4 pCi/L. The overall average of the fifty houses was some two to three times the United States value as well. The response to her efforts was one of outcry, concern, frustration, encouragement, and all gamuts of emotion. Many homeowners were angry, but were unable to direct their anger. Others were concerned about property values, others about the lack of government regulation on radon, and so on.

Interestingly enough, it is now estimated that, due to indoor radon, about 100,000 people (or more) living in the United States are exposed to the same radiation levels as the 100,000 people evacuated from the environs of the Chernobyl nuclear power plant during the 1986 disaster in the Soviet Union. The difference, though, is that in the latter case the presence of high amounts of radiation was known to exist and measurements and evacuations followed. In the case of indoor radon, however, most people probably do not know they are actually exposed to very high levels.

Recently, WJLA-TV, Baskin's employer in Washington, D.C., sponsored a vigorous study of 74,000 homes in the metropolitan D.C. area using charcoal detectors (see chapter 6 for discussion of these detectors). One third, or about 25,000 homes, yielded radon in excess of 4 pCi/L. While the method used is valuable only for screening purposes, it does emphasize the potential for high radon levels in indoor air.

After some introductory material on units, standards, and related topics, I will cover such matters as where and how uranium (the parent to radon) occurs in rocks, how radon is formed, how radon enters buildings from soil and water, as well as natural background radon in soil, water and air. In later chapters, I will discuss radon studies in the United States and other countries, methods for detecting radon, remedial steps to achieve lower radon levels, health effects of high radon levels, practical use of radon for economic prospecting, and radon risk relative to other commonly encountered risks and will offer some conclusions and recommendations.

This book is not intended as an exhaustive tretment of all radon work. For the more technical studies the reader is referred to the books listed in the appendix and many articles published in the scientific literature, especially *Health Physics*.

Notation

In this book, I will use the standard nomenclature for the elements and isotopes, as well as the proper chemical symbols for all aqueous species, minerals, and other compounds or species discussed.

I have already introduced isotopes of radon and of uranium. Thus, radon-222 is more properly written as ^{222}Rn and uranium-238 as ^{238}U and so on. I will follow this practice throughout the remainder of the book.

For minerals, I will often use the chemical formulas. Thus, the mineral uraninite, which is uranium dioxide, is written as UO_2. Species in solution may vary considerably when written. Radium is often transported as a simple ion with a positive charge of 2 and is written as Ra^{2+}. Some ions are very complex; an important ion of dissolved uranium in the presence of dissolved carbon dioxide is known as uranyl dicarbonate ion, or $UO_2(CO_3)_2{}^{2-}$ which tells us that this ion contains uranium and carbonate and has a negative charge of 2. All minerals and species mentioned will be tabulated in the glossary.

Radiation Doses and Background

As we will see in chapter 1, the dose of radiation to humans and other animals is usually given in rems (R) or thousandths of a rem, millirems (mR). To put these terms into perspective, normal background radiation in the United States at sea level is about 100 mR per year—without considering indoor radon. Due to cosmic rays from the sun, values above 100 mR are found with increased elevation above sea level. Similarly, many other factors, some manmade and some natural, contribute to our radiation background. These are covered in chapter 11 in greater detail.

The contribution of indoor radon to natural background radiation is fairly new. Until just a few years ago, it was not considered a factor. Now (see chapter 1) it is felt that, on the average, some 40 percent of background radiation may be from indoor radon. This necessitates revision of many standards, safety levels, etc., for background radiation, but that topic is outside the scope of this book.

This book is divided into twelve chapters plus an introduction, which I will briefly describe here. Chapter 1 deals with units that are used to describe radon levels as well as some other aspects of our natural radiation background and environment. It also deals with standards, the problems associated with the determination of such standards, and related areas. Special problems of both units and standards are also discussed.

Chapter 2 is a fairly detailed look at uranium and thorium, as well as radium, in the earth's crust. It is a lengthy, but necessary, chapter since uranium and thorium are the parents to all the earth's radon. This chapter discusses uranium and thorium chemistry, but in a digestible fashion. Some information on uranium ore deposits is included as well.

Chapter 3 describes the actual decay of uranium and thorium, with emphasis on ^{238}U, the parent to ^{222}Rn. I focus on the progeny of ^{222}Rn, for it is these radon daughters that cause the most damage to lungs.

Chapter 4 discusses in some detail how radon and radon daughters affect lung tissue and how data from former miners are used to determine the number of probable lung cancers for lower doses of radon and its daughters.

Chapter 5 takes a long look at how radon enters dwellings, including discussion of soil, building materials, air, waters, and other sources. Different factors such as weather are also included here.

Chapter 6 is concerned with the different methods used for the determination of radon levels. In this chapter, I describe the methods in lay terms and discuss the advantages and disadvantages of each.

Chapter 7 briefly covers the radon studies conducted—including ongoing studies—in the United States; there is, however, an uneven effort to study radon in the nation. The role of the various federal agencies as well as some states is examined. Finally, some information on the positions of the builders and realtors is covered, along with commentary on the role each may have to play in the future.

Chapter 8 deals with some of the common methods of remedial action that can mitigate high radon levels. These include the entirely passive approaches, such as the use of simple window frames, to more elaborate methods using pipes in walls and floors, under foundation slabs or basement floors—all with gas-pumping units, which are recommended by Environmental Protection Agency (EPA) or others. This chapter is generic, but some case histories are included.

Chapter 9 covers some aspects of radon studies on the international scene, including the rather extensive studies done in Sweden and in some other European countries, the large studies performed in Canada, and the few studies done elsewhere in the world. What will be very apparent from chapter 9 is that radon studies in developed countries are far more numerous and extensive compared to the paucity of such studies (if any at all) in the third world countries.

Chapter 10 deals with the various uses of natural radon emanations. Geologists, and especially explorationists, have taken advantage of the fact that radon emanations may be used to find certain kinds of ore deposits, to locate faults, to explore for petroleum and natural gas, to investigate areas for geothermal potential, and for a wide variety of other actual and potential economic purposes.

Chapter 11 is a brief look at some of the common sources of risk in today's world with focus on the United States. Here it will be seen that indoor radon becomes one of the largest natural causes of fatalities, compared in sheer numbers with fatalities from many anthropogenic sources.

Finally, chapter 12 will attempt a brief summary of the indoor radon problem. There I will draw some conclusions and make some recommendations for future work on this pressing and very important topic.

1

Radon Units, Standards, and Related Items

The various ways of reporting radon data are often confusing, in part due to the many assumptions involved in obtaining the data and in part due to the different units. The feeling that even for a known radon level one can't be sure how safe or how dangerous this level is often contributes to the confusion. Further, this confusion opens the doors for unreputable individuals or groups to bilk the public. In this chapter, I will attempt to explain the units, the standards proposed or in actual use, and some of the scams that are being perpetuated on the public in the United States.

UNITS

In this book I will use the CURIE as the basic unit for radioactivity. Rigorously, one curie (with the symbol Ci) equals 37 billion (3.7×10^{10}) disintegrations per second of radioactive material. I choose the curie (Ci) instead of the BECQUEREL (Bq; defined as one disintegration per second) because most radon data are reported in curies (or fractions thereof) and not becquerels.

The amount of radioactivity given off by radon is extremely small and is measured in trillionths of one Ci (10^{-12}), which is usually re-

13

TABLE 1.1
Prefixes for Units

Factor	Prefix	Symbol	Factor	Prefix	Symbol
10^{18}	exa	E	10^{-1}	deci	d
10^{15}	peta	P	10^{-2}	centi	c
10^{12}	tera	T	10^{-3}	milli	m
10^{9}	giga	G	10^{-6}	micro	u
10^{6}	mega	M	10^{-9}	nano	n
10^{3}	kilo	K	10^{-12}	pico	p
10^{2}	hecto	h	10^{-15}	femto	f
10^{1}	deka	da	10^{-18}	atto	a

ported in PICOCURIES (with the symbol pCi), one picocurie equaling one trillionth of a curie. Other Latin prefixes for different quantities are given in table 1.1; the reader may encounter all of these in today's literature about radon.

Again, for convenience, the amount of radioactivity of radon is reported for one liter of air (L). Thus, if two picocuries of radon were measured in one liter of air, it would be reported as 2 pCi/L. The ways in which radon is actually measured are discussed in chapter 6 and will not be covered now.

The radioactivity of radon is proportional to the amount of radon present. If high amounts of radon are encountered, the radon radioactivity measured per liter of air is high; if low amounts are present, then low radioactivity is detected.

The radon daughters, discussed very briefly in the introduction and in much more detail in chapters 3 and 4, are the main causes of lung cancer. The usual assumption is that for every amount of radon gas present, there is also present in the air some fixed amount of radon daughters. Thus, by measuring the radon content, one also has a way to measure the radon daughters that cause adverse health effects. Unfortunately, though, the behavior of the radon daughters is often different than that of the parent radon. This is because radon is a gas and the radon daughters are solids. The solids may be readily fixed by dust or aerosols or possibly even react chemically with other species in the air.

One way to deal with this problem is to use WORKING LEVEL (WL) as a unit, which is defined as the total radon daughters in one liter of air that will result in the ultimate emission of 1.3×10^5 MeV (million electron volts) of energy from alpha particles. In brief, if the radon is in equilibrium with the radon daughters, then the WL is equal to 100 pCiL of radon gas. Thus, 0.5 WL = 50 pCi/L. This WL unit is cumbersome, but it is widely used by the health workers in recognition that it is the radon daughters, not the radon gas, that is of concern to public health. If one wanted to be entirely rigorous, the amount of each radon daughter would have to be determined, but this is neither economical nor practical, so the WL is used. In addition, it is infinitely easier to measure radon than radon daughters, so often the radon daughters present are estimated from knowledge of the measured radon gas levels with the assumption stated earlier that 100 pCi/L radon radioactivity equals one WL. The National Academy of Science-National Research Council (NAS-NRC) and others recommend use of 200 pCi/L = 1 WL because of a number of observations that many radon daughters have been fixed on dust and are therefore not as easily taken into the lungs as radon.

When data are reported for radon, it may or may not be safe to assume a fixed amount of radon daughters based on this measurement. Yet, despite this uncertainty, the radon measurement gives a reliable, quick, and useful index for the safety of any dwelling in question. In mining and other industries, another unit, the working level month (WLM), is used. It is defined as one WLM = 1 WL for a 170 hour exposure. The EPA and others use WL and WLM extensively, yet most radon data in the United States, if not worldwide, are reported in pCi/L, as I will in the rest of the book.

The EPA recommends 4 pCi/L (0.02 WL) as a safe level for indoor radon, yet it is quick and careful to point out that there is no truly safe level of radon (see below and table 1.2).

Regardless of units, the common sense observation that higher radon levels are more dangerous than low radon levels still holds. We must learn to adjust to the indoor radon problem. We cannot legislate or regulate the earth. All of us must consciously make decisions about the radon levels around us.

STANDARDS FOR RADON

Everyone loves standards, and here standards are used in the sense of a level for safety. With standards, we think we know if some level of a potentially dangerous substance poses a threat or not. Yet standards are often in response to a lay public outcry fed most often by the media, and based on the need to set some value as a "standard," whether or not the science actually supports it. When first set, many standards are not scientifically sound; rather, they are working standards that must be examined critically. Most standards are set conservatively, so that by adhering to them risk to public health is minimized. Let me use carbon monoxide (CO) as an example. In Albuquerque, New Mexico, during the winter months, the standard for carbon monoxide is given as 100 parts per million (ppm). Above this level, one should neither jog nor do other exercise out of doors; people, especially the elderly, with lung disease are encouraged to stay inside their homes. Yet below 100 ppm CO is considered safe. So then, how much safer is 99 ppm than 101 ppm? Obviously, the difference here is nonsense. So too may be the setting of a standard for something as enigmatic as radon. While the EPA may recommend 4 pCi/L as a working standard, someone with lungs more susceptible to damage from radon daughters, such as the very young or the very old or those with weakened lungs, may be at a much higher risk than a young adult in excellent health.

I will refer to this 4 pCi/L value because it is the most widely used figure in the United States, but the reader is cautioned that there may be significant risk even at this level. In the introduction I mentioned that the 4 pCi/L figure was obtained based on a study of noncontaminated (i.e., with uranium mill tailings as building materials) houses in Grand Junction, Colorado.

A further complication of background radiation arises from the fact that standards for exposures to man-made radiation have been set without regard to the radiation from indoor radon. The International Council on Radiation Protection (ICRP) and the National Council on Radiation Protection (NCRP) both adhere to an exposure limit per individual in the United States from manmade sources of 500 millirems (mR) and recommend that any group within the general population must not be exposed to more than 170 mR. Yet, as will be pointed out in some detail in chapter 11, the natural background radiation from indoor radon may be several hundred mR in some cases, thus outweighing the potential contributions from man-made sources.

The EPA also recommends remedial action to be taken for dwellings with 4 pCi/L or above (see table 1.2), although the ICRP uses 8 pCi/L before remedial action is recommended.

TABLE 1.2
Estimated Lung Cancer Deaths[a]

Indoor Radon Level (pCi/L air)	Deaths from Lung Cancer per 100 People
4 pCi/L	Between 1 and 5 for every 100 individuals
20 pCi/L	Between 6 and 21 for every 100 individuals
200 pCi/L	Between 44 and 77 for every 100 individuals[b]

[a] This assumes 70 years in the dwelling at 70–80 percent of time indoors.
[b] If an individual lived in this dwelling for 10 years at this high (200 pCi/L) indoor radon level, his or her expected risks would be between 14 and 42 deaths per 100.

How many homes in the United States exceed 4 pCi/L? This topic is covered in detail in chapter 7. Here, however, let me note that A. V. Nero, Jr. (1986, 1987), argues that seven percent of the single family homes in the United States exceed 4 pCi/L and two percent exceed 8 pCi/L. Others (Alter and Oswald, 1987) also argue for 20 and 30 percent of home rate over 4 pCi/L. In the Albuquerque, New Mexico, area, my studies show these figures to be 28 and 12 percent respectively.

The EPA (1986) is careful to emphasize that the urgency of remedial action depends on the level of indoor radon detected. The higher the level, the greater the urgency. Remedial methods to lower indoor radon levels are described in chapter 8.

Very strict standards for new houses have been suggested by the American Society of Heating, Refrigerating, and Air-Conditioning Engineers (ASHRAE), who recommend 1 pCi/L air for radon daughters. ASHRAE concern is with total indoor pollutants, not just radon, however.

Also with reference to radon daughters, the World Health Organization (WHO) argues for about 10 pCi/L for ordinary existing buildings and about 3 pCi/L for future buildings.

Internationally, Canada favors 2 pCi/L for radon daughters in contaminated buildings, while the United States argues for a range of 1 to 5 pCi/L, depending on the magnitude of the contamination. In the Fennoscandian countries (Denmark, Finland, Iceland, Norway, and Sweden), 10.8 pCi/L for radon daughters is given for ordinary buildings and 2.7 for future buildings.

Just how many fatalities from indoor radon are there each year? For the United States, the EPA and NRC have given figures from 8,000 to

30,000, while the prestigious National Academy of Science (NAS) has recently (1988) given a figure of 15,000 lung cancer fatalities per year. A figure of 15,000 is very large, but still only about one tenth of the lung cancer fatalities from cigarette smoking.

WHAT'S SAFE? AND WHAT ARE SOME OF THE PROBLEMS INVOLVED IN DETERMINING SAFETY?

If one assumes a lifetime of approximately 70 years, living in the same dwelling 75 percent of the time, then it is possible to estimate the risk of lung cancer fatalities from indoor radon (see table 1.3). These figures are not "statistically" influenced; they are based on careful extrapolations of detailed studies of those who mine uranium and other metals, as well as others whose occupations expose them to radon and/or radiation.

TABLE 1.3
EPA Recommendations for Remedial Action

Results higher than 200 pCi/L "Exposures in this range are among the highest observed in homes. Residents should undertake action to reduce levels as far below 200 pCi/L as possible. We recommend that you take action within several weeks. If this is not possible, you should determine, in consultation with appropriate state or local health or radiation protection officials, if temporary relocation is appropriate until the levels can be reduced."

Results from 20 to 200 pCi/L "Exposures in this range are considered above average for residential structures. You should undertake action to reduce levels as far below 20 pCi/L as possible. We recommend that you take action within several months."

Results from 4 to 20 pCi/L "Exposures in this range are considered above average for residential structures. You should undertake action to lower levels to about 4 pCi/L or below. We recommend that you take action within a few years, sooner if levels are at the upper end of this range."

Results about 4 pCi/L or lower "Exposures in this range are considered average or slightly above average for residential structures. Although exposures in this range do present some risk of lung cancer, reductions of levels this low may be difficult, and sometimes impossible, to achieve."

Source: Environmental Protection Agency 1986.

However, there is no international agreement as to what the radon standard should be, although most nations do agree that lowering the radon levels to "acceptable limits of risk" is desirable. Since the objective of any standard is to promote better health, in the case of a standard for radon, the intent is to reduce the number of lung cancer fatalities presumably due to this gas. But how is this standard to be implemented? Should it be applied across the board to the entire population of a country or, as data become available, to areas where there is a known incidence of high radon (such as the Reading Prong of the northeastern United States)?

If remedial measures are focused on those dwellings with high indoor radon values, say above 20 pCi/L, then there is the risk that owners of the larger number of dwellings in the range 4 to 20 pCi/L will have no incentive to lower the radon levels.

There are several basic problems involved here. First, there is a need in any country, not just the United States, for quality radon surveys of as many representative homes as possible. Yet, at the same time, it is unrealistic to suppose that more than a small fraction of the dwellings of the total population of the United States will be tested for indoor radon levels. To overcome this problem, there must be an initial and directed plan to selectively sample in areas of possible high radon while at the same time encourage reconnaissance indoor radon detection in the rest of the areas. There are many problems involved. How are the data taken, as short-term, large-error data or more reliable, long-duration data? Assuming that the data are reasonable, how proprietary are they? Do homeowners have to make public the results of the radon studies of their homes?

If the EPA or some other federal or state agency gathers the data and if such data turn out to be above 4 pCi/L, can the home owner prevent the data from becoming public? Further, should a home owner know that his or her dwelling's radon level is high, is he or she required to advise perspective buyers of this fact? Similarly, are realtors who are involved responsible?

Still further, if a building is new and later found to have high indoor radon, who is liable for any remedial action or possible legal actions resulting from an adverse health consequence presumed due to indoor radon? If a prospective buyer insists on a radon measurement, what guarantees are there that the detector (and most are very small) will be left in place during the entire test period (i.e., cannot someone can place the detector out of doors for a part of the test period)?

And who rides herd on the persons or groups doing the radon measurements and remedial work? At present, the EPA will list those companies or individuals who participate in their round-robin testing of radon measuring apparatus. This is done by the firms sending their testing apparatus to the EPA, which analyzes a known radon concentration. If the apparatus passes this test, then the firm is listed by the EPA, and this information is available to the public. I strongly advocate that consumers take this step; stories of flagrant fraud in the radon-measuring and remedial-action work abound, and opportunists are already fleecing the public in many areas (at least in the United States).

One thing appears certain at this time: Without realistic standards, the means to enforce them, and some legal description of responsibility for radon levels, there will be a huge onslaught of legal battles. Recently, Kirsch (1986) reported that a family whose home contained high radon levels sued the contractors on the grounds that they had to vacate the home due to the radon and, further, that possible chromosomal damage has resulted from exposure to the radon. Since over a short time there is probably no way to document this positively or negatively, the outcome is in the hands of the lawyers. Most people bringing suit in environmental matters do so in civil court where a unanimous jury vote is not required, just a simple majority vote. Again, the talents of a clever lawyer may be much more important than any actual radon level.

I make the point again: Without standards and policy, regulations and education the radon scare and concomitant opportunism will accelerate.

WHAT IF I SMOKE?

The question of smoking and indoor radon is an interesting one. Most probably there is a multiplacative effect, i.e., if you smoke and are exposed to indoor radon, your chances of fatal lung cancer are higher than they are from either source alone. Yet data are confusing and not clear in this area. Cohen (1979) presented early data suggesting that cancers caused from radon were different from those caused by smoking (see discussion in chapter 4), yet the recent study of the NAS (1988) presents new data that do not support any clear distinction between lung cancers from radon and those from smoking. They point out that of the studies involving former miners, the smoking habits of the individuals studied, such as type of tobacco, amount smoked, inhalation experience, cessation of practice, stop-start, and so on, were not always available.

However, smoke in the air is an excellent media on which radon daughters and radon can become attached. An interesting question is that of radon-exacerbated lung cancers caused by passive smoking. Already it is estimated that 5000 individuals in the United States die each year of lung cancer from passive smoking (see discussion in chapter 11). No data, unfortunately, are available on the number of deaths of non-smokers caused by a multiplacative passive smoke-indoor radon connection.

The implication from this discussion, of course, is that the actual number of lung cancer fatalities in which radon plays an important part is probably far greater than the 15,000 figure presented by the National Academy of Science. Of the total 150,00 lung cancers due to smoking each year, radon no doubt plays a significant role for many but, unfortunately, ways to document just how many are not yet available.

2

Uranium and Thorium in the Earth's Crust

Uranium and thorium are the ultimate parents of all radon found in the earth. How these elements are distributed in rocks, minerals, soils, and waters is therefore of interest and relevance to the distribution of radon in the earth as well. The emphasis in this chapter is on uranium, but thorium is included as well. In addition, some brief discussion of uranium ore deposits is also included because of its importance for potential radon release to the atmosphere and to illustrate the use of radon in search for such deposits.

Some years ago, the U.S. government made an intensive search for uranium deposits under its National Uranium Resource Evaluation (NURE) program. Now NURE offers a guide for specific areas for radon testing. For this reason, I will also include some discussion of the NURE program.

CHEMISTRY OF URANIUM, THORIUM, AND RADON

Uranium is a trace element in the earth's crust, but sometimes it becomes enriched to form ore deposits. It occurs in rocks and minerals as an ION with a charge of +4, which we write as U^{4+}. If, however, this uranium should come into contact with waters containing dissolved oxygen, then it will be oxidized to another ion of uranium, but this time with a charge of $+6$, or U^{6+}. The difference between the two ions is important for the discussion of the uranium behavior in rocks, soils, and waters of the earth's crust. On the one hand, the $4+$ ion is highly insoluble; if released from a rock or mineral, it will combine with water and form uranium hydroxide, $U(OH)_4$, which is immobilized immediately. The $6+$ variety, on the other hand, is extremely soluble. It forms various complex ions with oxygen and with carbonate. Since carbonate is present in most surface waters, the uranium carbonate complex is how uranium is most commonly transported in nature.

This different behavior of the uranium ions is very important for the distribution of radon. Should uranium be fixed as insoluble $4+$ minerals, then the source for radon is, by definition, fixed in this spot also. Conversely, should the uranium be $6+$, it will be transported away from the source rocks and either accumulate in waters or be fixed in soils away from the source.

It is also important to know the existence of oxidation-reduction reactions in nature. OXIDATION means that a reaction occurs so that electrons are freed; for example, the oxidation of native iron to red iron oxide, with iron having a charge of $+3$, is rust. In this reaction, the iron is said to have been oxidized; i.e., the charge on the ion increases from zero (native iron) to $+3$ (in the rust). The energy for this oxidation reaction usually comes from dissolved oxygen. The opposite of oxidation is REDUCTION. If the rust were to be deeply buried in the earth so that no more oxygen were available, then, over time, the rust would be transformed back to native iron. When the charge on an ion such as iron is reduced from $+3$ to 0, the reaction is called reduction.

Oxidation and reduction reactions are all around us. Iron that occurs in many minerals has a charge of $+2$, intermediate between iron and rust. This iron with $+2$ charge will, like native iron, also oxidize to iron with a charge of $+3$. Now why is this important for radon?

Rocks in which iron $+2$ has been oxidized to iron $+3$ are easy to spot because the iron $+3$ minerals such as hematite, goethite, and others are bright yellow, brown-red, and red. We have all noted the deep reddish soil in the tropics or on weathered rocks in arid climates. In all these rocks, iron $+3$ has been formed, and, because it is insoluble, the brightly colored iron minerals form there on the surface and the oxidation may be so severe (as it is in the tropics) that eventually all iron $+2$ is converted to iron $+3$.

The oxidation of uranium $+4$, the insoluble variety, to uranium $+6$, the soluble variety, takes place at a lower energy than that required for iron $+2$ to be oxidized to iron $+3$. Thus if the rocks are colored reds and yellows because of the iron oxidation, then we know that any uranium along grain edges or other surfaces has been oxidized to the more soluble $+6$ variety. This simply means that these rocks may have lost some of their uranium. This is not at all uncommon in the arid southwestern United States, for example. But how can we unequivocally tell if such weathered rocks have lost their uranium, and, if so, how much?

To evaluate these questions, we must consider the element thorium for a minute. Thorium, like insoluble uranium, occurs in nature as a $+4$ ion in rocks and minerals. Unlike uranium, though, thorium contains only a $+4$ charge even under the highest conditions of oxidation. Thus, under conditions where iron and uranium are both oxidized, the thorium stays as $+4$ thorium. Further, like uranium $+4$ minerals, the thorium $+4$ minerals are all highly insoluble under most earth conditions. Now the first question can be answered based on this observation—coupled with the fact that the ratio of thorium to uranium for the whole earth's crust is about 4. If samples are found, for example, that contain very high ratios of thorium to uranium, say about 20 to 80 or higher, then the chances are that these rocks have lost much of their uranium.

The "how much" is more difficult to answer and can only be addressed by looking carefully at the exact budget of the uranium decay products (fig. 2.1). It is not difficult to imagine a scenario in which a rock can gain uranium, just as some have lost uranium. Imagine a rock that loses its uranium but retains its thorium. This soluble uranium is removed in groundwaters. At some point, these waters, which are oxidizing, penetrate deeper and deeper into the earth so that the overall chemical conditions become reducing. When this happens, all the uranium brought in as $+6$ is changed to uranium $+4$ and precipitated. This is how some uranium ores form, for example. These uranium-rich rocks are very deficient in thorium, because only uranium has been carried by the water's reaching the zone of uranium mineral precipitation.

FIGURE 2.1

Cross section of an hypothetical uranium ore roll deposit (See text for details), showing position of former ore roll, and distribution of uranium (U), carbonaceous matter (C), vanadium (V), selenium (Se) and molybdenum (Mo). The remnant ore, as well as radium fixed in the rocks near the position of the former roll site, may be sources of radon emission.

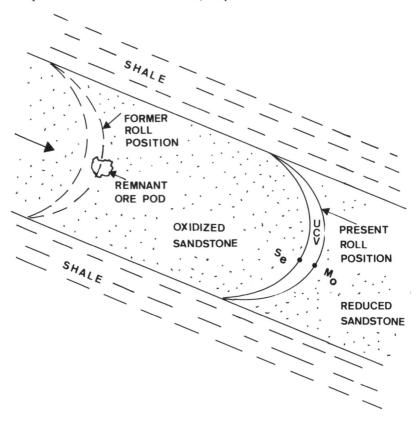

Let me put this into perspective for radon. First, it is important to remember that even if water that comes into contact with rocks is oxidizing, this water can readily attack only the surface of minerals, so most of the uranium present is not affected. This must be the case, otherwise one would never be able to date a rock by the so-called uranium-lead methods described later in this chapter—methods that demand that the parent (uranium) and all the daughters of radioactive decay (all the isotopes shown in fig. 2.1, for example) remain essentially 100 percent in the rocks. Yet "essentially 100 percent" is usually not quite 100 percent, and this slight deviation is due primarily to the loss of some of the intermediate radioactive decay products of uranium such as ^{222}Ra.

Obviously the chemistry of uranium is pretty complex. In order to use our knowledge of uranium to help us address the radon problem, it is necessary to know something about the uranium distribution in rocks as well.

ROCKS

Igneous Rocks. Igneous rocks form from molten rock called **MAGMA**, which, in turn, is called **LAVA** if molten rock erupts from volcanos at the earth's surface. Granite and basalt are two familiar igneous rocks.

Magmas mainly consist of the dissolved major constituents of rocks such as silica, aluminum, iron, potassium, magnesium, calcium, and sodium. Silica (SiO_2), the most abundant constituent, is used to distinguish between different types of igneous rocks. Granites contain high amounts of silica and are said to be silica-rich, or acidic. Basalts contain much less silica and are said to be silica-poor, or basic. Uranium as a $+4$ ion has difficulty entering the structures of most rock-forming minerals and tends to be segregated into the last liquid to crystallize from the magma. Thus, it is often fixed in minor minerals called accessories. Further, uranium is more enriched in acidic (silica-rich) magmas than basic (silica-poor) magmas. Even in the silica-rich magmas, some of the uranium forms discrete little crystals in between the major and accessory rock-forming minerals, perhaps as much as 30 percent of the uranium may be so fixed. Because of its location along grain surfaces and in cracks, etc., this uranium is especially vulnerable to leaching by groundwaters.

In table 2.1, the average concentrations of uranium and thorium are given for magmatic rocks for comparison with various sedimentary rocks. In table 2.2, the uranium abundances in a variety of igneous rocks are given. It is noted that the more basic rocks (basalt mainly) and the very silica-poor, or ultrabasic, rocks contain only one part per million (ppm) or so of uranium, whereas the more silica-rich rocks, from diorite-quartz diorite, through granodiorite to granite, contain two to five ppm.

TABLE 2.1
Comparative Concentrations (ppm) for Selected Elements

Upper Crust Element	Mag. Rx	Grywck.	Shale	Sandstone	Carbonates
K	28,200	17,800	30,000	10,700	2,700
Ca	28,700	17,000	15,800	39,100	(30%)
Rb	120	180	140	60	3
Sr	290	200	300	20	610
Pb	15	n.d.	20	7	9
U	3.5	1.8	3.7	0.5	2.2
Th	11	6.3	12	(2)	(1.7)
Zr	160	450	160	(200±)	—
V	95	67	130	(15)	var.
Se	0.09	n.d.	0.6	?	?
Mo	1	n.d.	1.6	?	?

Source: Wedepohl 1967.

While these differences may not seem like much, it must be remembered that there are considerable masses of these rocks in the earth's crust; so, even a slight loss of uranium from a granite is much more significant than a similar loss from basalt. Finally, in table 2.3, the uranium concentrations in accessory minerals are compared with the major rock-form-

TABLE 2.2
Uranium Concentration in Igneous Rocks (in ppm)

	Mean	Range
1. Ultramafic		
a) dunites, olivine nodules, etc.	—	.003–.05
b) pyroxenitites	0.7	——
c) eclogites (all)	—	.013–.80
2. Mafic igneous rocks	0.9	<.2–3.4
3. Diorite-quartz diorite	2	<.5–11.5
4. Granodiorite	2.6	<1–9.0
5. Granites	—	2.2–15.0
6. Other silicic igneous rocks	4.7	<1–21.0
7. Other silicic extrusive rocks	5.0	——
8. Alkali instusive rocks	—	.04–19.7

Sources: Clark et al. 1966; Rich et al. 1976; Rogers and Adams 1967.

TABLE 2.3

Igneous Minerals and Concentration of Uranium

Majors	ppm	Type of Occurrence in Mineral
Quartz	0.1–10	grain boundaries or crystal defects
Feldspar	0.1–10	?
Muscovite	2–8	?
Biotite	1–60	zircon inclusions?
Hornblend	0.2–60	zircon, biotite inclusions?
Pyroxenes	0.01–50	zircon, biotite inclusions?
Olivine	0.05	?
Accessory Minerals		
Allanite	30–1000	ionic substitution
Apatite	55–150	
Epidotre	20–200	
Garnet	6–30	
Ilmenite	1–50	
Magnetite	1–30	
Monazite	500–3000	
Sphene	10–700	
Zenotime	300–35,000	
Zircon	10^2–6×10^3	

Sources: Clark et al. 1966; Rich et al. 1976; Rogers and Adams 1967.

ing minerals. The major rock-forming minerals commonly contain less than one ppm (i.e., one part per million, or 0.0001, weight percent), although inclusions of very tiny crystals of accessory minerals such as zircon may give locally high values. Most of the accessory minerals, however, contain very large amounts of uranium. Zircon may contain up to 6000 ppm (0.6 weight percent).

It must also be emphasized, though, that perhaps as much as 30 percent, of uranium does not even enter the common accessory minerals, but rather is fixed along grain interstices and in crystal defects, etc. Again, this uranium is especially vulnerable to attack by waters and can be readily leached. If this kind of leaching has occurred, the rock as a whole will have a very high ratio of thorium to uranium and thus be easy to spot. I will return to the importance of this in the discussion of project NURE later in this chapter.

Sedimentary Rocks. Uranium is widely and sporadically distributed in sedimentary rocks, mainly because of the different sources of uranium, the different fixation mechanisms, and related factors. Table 2.4 summarizes the uranium data for several common sedimentary rock types, giving the ranges of uranium determinations and, where appropriate, the arithmetic mean.

Shales commonly contain organic matter, material that is very efficient in causing uranium reduction and fixation. Hence, shales usually contain uranium in higher amounts than in other sedimentary rocks, and in some black shales, such as the Alum Shale in Sweden and the Chattanooga Shale in the eastern United States, the uranium contents may be in the hundreds to perhaps even thousands of ppm. The problems associated with the Alum Shale and indoor radon are mentioned in the introduction and chapter 9.

Sandstones normally contain little uranium. If, however, the sandstones contain admixed shale, and especially shale plus organic matter, these admixed sites are favorite fixation places for uranium to be readily transported through the highly permeable sandstone. A **WACKE**, which is a very clay-rich sandstone, will then typically have higher uranium than a clean, clay-poor sandstone. Sedimentary quartzite, or orthoquartzites, contain just the mineral quartz. Table 2.3 indicates that quartz does not contain much uranium; hence, the orthoquartzites cannot either.

In carbonate rocks, such as limestone and dolomite, uranium is moderately abundant, although thorium is not. Uranium is, as mentioned earlier, commonly carried as a uranium carbonate ion in solution. As carbonate minerals form from sea water, some uranium is incorporated into the rocks, such as limestone, with relative ease. Obviously, the source rocks for the uranium will be very important here. For a limestone formed in waters near weathering granite source rocks, the uranium content may be quite high. If the source rocks being weathered are low-uranium rocks, such as orthoquartzites or evaporites, then the limestones are likely to be uranium-poor. Most limestones form in shallow seas with considerable input from the surrounding continental weathering rocks.

There are two kinds of evaporites; both form by the evaporation of trapped waters. On the one hand, marine evaporites form by evaporation of sea water that is trapped in some kind of structure on the edges of continents, and the resultant chemistry and mineralogy of the salts formed by evaporation must reflect the chemistry of the seas. Nonmarine evaporites, on the other hand, form in isolated basins not in contact with the oceans. Great Salt Lake, for example, if evaporated, would yield a nonmarine evaporite. Hudson Bay, which is in communication with the oceans, would yield a marine evaporite. There is not much uranium in the oceans. Hence, when marine evaporites form, they are especially depleted in uranium, as seen in table 2.4. Some nonmarine evaporites, however, and especially those with admixed organics (although not shown in table 2.4), may contain appreciable uranium if the source rocks feeding the isolated basins are uranium-rich.

Marine phosphorites are shales rich in the mineral apatite, a calcium phosphate that also contains appreciable uranium. These rocks also present many problems. The uranium in them is a source of radon, and

TABLE 2.4
Summary of Uranium Data for Various Sedimentary Rocks (in ppm)

	Arithmetic Mean	Range
Fine-grained clastics		
common shales	3.5	1–13
North American gray and green shales	3.2	1.2–12
Mancos shale (New Mexico)	4.2	0.9–15
Black shales	8.2	3–250
Coarse-grained clastics		
sandstone (arkose & graywacke)	—	0.45–3.2
orthoquartzites	0.45	0.2–0.6
Carbonates		
carbonate rocks	2.2	0.1–9
Russian carbonates	2.1	—
North American carbonates	2.2	0.65–8.8
California limestone	1.3	0.03–4.9
Florid limestone	2.0	0.5–6
Other sedimentary rocks		
marine phosphorites	—	50–300
evaporites	—	0.01–0.43
bentonites	5.0	0.1–21
bauxites	8.0	3–27

Sources: Clark et al. 1966; Rich et al. 1976; Brookins 1979.

phosphorite wastes upon which homes have been built may be prone to high radon from the uranium fixed there. The phosphorites also contain fluorine, which is also a pollutant to the environment and which may also help to move uranium about as a soluble uranium fluoride species. For many years, the high uranium content of the phosphorites was ignored. Now, the uranium is recovered as a byproduct of phosphate mining in Florida, but prior to the establishment of the mills, the uranium wound up with the wastes dumped into the surrounding swamps or used for fill. Further, the waste gypsum from the phosphate ores is used for wallboard, and this material, too, contains above background amounts of uranium. Thus, this gypboard may also be a source for radon. The state of Florida has now undertaken a rigorous program to investigate in detail the radon impact of the phosphorite wastes.

Bentonites are formed by weathering of volcanic rocks and may, if the source rocks are uraniferous, also contain appreciable uranium. Their mineralogy and chemistry is very complex, and uranium analysis is the only way to predict which bentonites will be uranium-rich and therefore radon-rich.

Bauxites and other laterites are formed under conditions of extreme tropical weathering—torrential downpours, very rapid oxidation, rapid runoff, high temperatures, and so on. Much of the uranium released is trapped in the newly forming aluminum and iron hydroxides that make up these rocks; hence, bauxites generally have relatively high uranium contents.

Metamorphic Rocks. Uranium abundance in metamorphic rocks is highly variable. The range of uranium concentration in these rocks is too wide to allow reasoned predictions. By processes commonly grouped together under the general heading of metamorphic differentiation/segregation, minerals tend to increase in grain size as a function of their increasing metamorphic grade (to minimize the surface area of minerals and, hence, the energy of the system). Trace elements tend to be removed from original minerals and redistributed into others or even removed from the rock. In short, uranium is not concentrated in metamorphic rocks unless it is there to begin with; alternatively, if a relatively uranium-rich rock is metamorphosed, very rich uranium veins can result.

TABLE 2.5
Summary of Uranium Concentrations in
Various Waters

Natural Waters	
Sea water	1–4 ppb
Continental waters	0.1–1000 ppb

Sources: Rich et al. 1976; Scott and Barker 1962.

WATERS

Uranium concentrations in waters are tabulated in tables 2.5 and 2.6. Typical sea water contains 1–4 ppb (1 ppb = .001 ppm), and continental waters range from 0.1 ppb (or less) to hundreds of ppb. Concentrations of 100 ppb (or more) generally denote mineralized areas. However, uranium concentrations vary widely, indicating that terrane classifications alone are not significant indicators of uranium deposits.

URANIUM DEPOSITS

Sandstone Uranium Deposits. The most frequent uranium deposit in the United States is that found in sandstone. In such a deposit, the

TABLE 2.6
Uranium in Waters over Specific Rock Types

Type	$\bar{x}$	Range	n(>4 ppb)	%(>4 ppb)
Igneous				
silicic	4.5	0–32	12	36
basic	0.9	0–9.2	1	6
Sedimentary				
sandstone and conglomerate	26	0–2000	22	17
silt and shale	11	0–69	6	43
limestone and dolomite	2	0–33	11	12
sand and gravel	2.5	0–74	13	15
Metamorphic rocks				
(all types)	4.4	0–37	8	24

Sources: Rich et al. 1976; Scott and Barker 1962.

uranium is concentrated in small zones rich in shale and organic matter in an otherwise highly permeable sandstone that is over- and underlain by relatively impermeable shales. Here the sandstone acts as a favorable aquifer, while the shales do not. Further, when the organic-rich shale zones are encountered in the sandstones, any uranium carried in solution (for example, uranium carbonate ion) will be precipitated if the oxidation potential of the system is adequate.

If the uranium concentration in solution is high, precipitation as a uranium 4+ mineral could occur above the so-called REDOX boundary. The redox boundary marks that point in the rocks where the chemical conditions change from dominantly oxidizing to dominantly reducing. Thus, above the redox boundary, uranium 6+ occurs; below it, uranium 4+ is found (also, iron 3 above, iron 2 below, etc.). Two common uranium 4+ minerals that form in the reducing condition rocks are uranium oxide, uraninite (U_3O_8), and uranium silicate, coffinite ($USiO_4$).

The uranium precipitation commonly occurs in the presence of organic matter. Thus, the assemblage uraninite (or coffinite)-organic carbon-pyrite is typical of these deposits. Elements such as vanadium, molybdenum, and selenium are commonly reduced at about the same time; hence, they are also associated in the deposits.

The classic picture of a sandstone uranium deposit is shown in figure 2.1. I have described the formation of such deposits. These occurrences, called roll front deposits, are usual for many young ores in the United States, especially in Wyoming and Texas. Their formation generally follows a characteristic pattern. When a layered shale-sandstone-shale sequence is tilted, ground waters move more readily through the sandstone layers than through the shale layers. In many cases, the original redox front of the formation was probably the water table. Regardless of this point, the redox front, the locus of reduction for many species, separates the deeper buried and reduced sandstone from the more shallow oxidized sandstone. This is the point at which the metals uranium, vanadium, molybdenum, and selenium are enriched.

The fronts of such roll deposits apparently move down the gradient as either the water table lowers or, more likely, as more and more oxygenated waters penetrate the permeable sandstone. Although the mechanisms for this movement are not exactly known, field evidence showing that small, remnant ore pods are left behind in oxidized ground suggests that frontal movement is downward. These ore rocks are, of course, very favorable source areas for radon emission.

Other Types of Uranium Deposits. Approximately 50 percent of the world's known uranium reserves are within Precambrian "conglomerate" deposits, and a lesser, but still significant, amount are within vein deposits. The Precambrian deposits of uranium are characterized by the following features: 1) Their uranium is found in association with sulfide minerals; 2) the deposits are conglomerates having abundant quartz and other detrital grains; and 3) the ore is structurally controlled in certain favored zones by nonconformities between underlying, often metamorphosed rocks, and overlying, frequently unmetamorphosed sandstone-rich rocks. The well-known deposits in the Northern Territory of Australia and many of those in northern Saskatchewan, Canada, are of this type. Although smaller and more difficult to locate than sandstone deposits, the Precambrian deposit usually has a very high uranium content (30 to 50 weight percent of uranium metal in some instances), and the aggregate uranium reserves of both types of deposits are comparable.

Less important than the Precambrian conglomerate deposits are vein deposits.

Except for small, uraninite-rich pegmatites, most igneous rock deposits have not received much research. Although the assumption was routinely made that the uranium in igneous rocks is too low grade and too difficult to remove to make such deposits an economical source of uranium, discovery of the Rossing Deposit in Namibia (formerly the Republic of Southwest Africa) has suggested otherwise. This deposit is a uranium-rich, very extensive body of granite—in fact, it has been referred to as a porphyry uranium deposit, i.e., analogous to porphyry copper deposits.

Precambrian vein deposits have been found in the United States. In Virginia, a long-range plan calls for mining of such a deposit. Discovery of this and other smaller deposits nearby attest to the fact that these rocks can contain considerable uranium. In turn, this means that higher radon levels can be expected in such areas. The Reading Prong (see introduction) contains many small uranium rich areas, which, while most are not ores, are very productive radon sources (see the Watras Case in the introduction).

The Chattanooga Shale along with other uranium-enriched shales have been listed as potential low-grade uranium resources, although metallurgical and/or environmental constraints pose problems. If these shales should be mined for their marine phosphate minerals, then, of necessity, the tailings will have to be treated for uranium removal. Where these shales underlie areas developed for housing, a high potential for radon emanations occurs.

The extensive deposits of indurated, $CaCO_3$–rich **CALICHE** in the Western United States commonly possess abundant uranium—but the deposits are small and sporadically distributed. Future technology may allow recovery of uranium from these deposits, but more importantly, the possibility of using the caliche deposits to trace uranium distribution in related rocks to see if enrichment, i.e., of the sandstone-type, has occurred is presently being explored. This uranium-rich caliche may often be foundation material for buildings in the southwestern United States, and thus a radon source may be built into the structure. Great care must be taken with not only the materials used, but also the location of the building.

Following World War II, the Atomic Energy Commission (AEC) established a long-range program to develop alternative sources of uranium other than those deposits characterized by the Colorado Plateau in the western United States. These alternative sources include uranium in phosphate rock, shale, monazite, and lignite. Of these, large-scale recovery of uranium from phosphate rock looks most promising. The U.S. Geological Survey (USGS), cooperating with the AEC, found that the highest concentrations of uranium in phosphates occur in Florida, Montana, Utah, Wyoming, Nevada, and Colorado. These phosphate reserves are estimated to contain as much as 600,000 tons of recoverable uranium, of which approximately 48 percent are in Florida. The radon potential of these rocks has been mentioned earlier.

The richest phosphate deposits in Florida are found principally in Polk, Hillsborough, Manatee, and Hardee counties in the west central portion of the state. The deposits are generally characterized by an overburden of Pleistocene quartz sand and **LEACHED-ZONE** material . The leached zone contains some phosphate, primarily in the form of aluminum phosphate minerals. Its formation was probably caused by weathering and groundwater leaching of the upper parts of the phosphate-bearing rocks. The principal phosphate deposits occur as calcium fluorapatite $[Ca_{10}(PO_4)_6F_2]$ matrix. The thickness of the matrix ranges from 0 to 50 feet and averages about 12 feet. It accounts for practically all of Florida phosphate production and about 80 percent domestic production.

Often the phosphate in the apatite structure can be partly replaced by the vanadate, silicate, sulfate, or carbonate ions. Rare earths, chromium, and uranium are also often common impurities. Any uranium present was deposited simultaneously with the phosphate rock and was incorporated in the grain structure. Thus, the uranium concentrations vary directly with the phosphate concentration of phosphate rock, and all this material is potentially high in radon.

METALLOGENIC PROVINCES EXPLORATION: IMPLICATIONS FOR RADON

The conceptual value of **METALLOGENIC PROVINCES** (as logical areas for uranium deposits) is debatable. At times one can question the sufficiency of data in areas that have been considered metallogenic provinces. Yet the concept is useful if one can correlate geologic criteria, chemical data, and other factors so that areas of high uranium, and thus potentially high radon, can be identified.

The Colorado Plateau meets many of the criteria required of a metallogenic province classification. The deposits therein can be correlated in terms of rock types, mineralization controls, chemistry, geological age, and other criteria. More importantly, the subeconomic (i.e., uranium content is above background but less than ore grade) areas have been studied in sufficient detail to be included in the province. Further, high radon is noted in most of the Colorado Plateau.

RADIOMETRIC AGE DETERMINATIONS

It is important to introduce here the concept of radiometric age determinations. The oldest attempts at dating rocks and minerals by radiometric means date back to the early part of the twentieth century, when B. B. Boltwood attempted to date uraninite (natural uranium oxide, UO_2) from a pegmatite in Connecticut. He assumed all uranium was ^{238}U (i.e., the isotope ^{235}U had not yet been discovered) and that all lead present was ^{206}Pb (the final, stable decay product of ^{238}U; see fig. 3.1). By assuming a rate of decay and no gain or loss of either uranium or lead (i.e., closed system conditions), he then calculated the age of the uraninite to be between 300 and 600 Ma (millions of years). He argued that the 300 Ma date was more reasonable because of the possibility of some lead present in the uraninite before decay began. Boltwood's work is important for many reasons, not the least of which is the fact that in order for this, and for any geochronologic method based on radioactive decay, to work, all the intermediate radioactive daughter products formed between parent uranium and eventual stable daughter, lead, must be 100 percent retained in the mineral or rock being studied. This is the most important premise on which radiometric age determinations are based, and geochronologists have been successfully dating rocks and minerals by the uranium-lead method since Boltwood's classic work.

For our purposes, this also tells us that most of the radon in earth materials is tightly locked into minerals such that it does not readily escape from the crystalline structure in which it is housed. Yet some radon does escape. Careful inspection of virtually all radiometric ages by the uranium-lead method shows that there has been some loss of intermediate daughters and/or lead. Although the amount of intermediate daughter loss is small, it nevertheless is a large figure for the earth as a whole. Further, when old rocks are deeply buried and melted, new magmas form that then crystallize into new igneous rocks, and all the radon built up prior to the melting is lost. Similarly, during strong prograde metamorphism of rocks, recrystallization takes place with, usually, loss of radon and other intermediate daughters. Further, because uranium minerals are altered by weathering, they can lose radon. All of these and other factors cause radon to be released into various earth materials, from where the radon can make its way to the earth's surface and thus into the earth's atmosphere—or into homes.

When uranium 4+ is oxidized to uranium 6+, this does not necessarily mean that it will be lost from rocks, but often some is. How far and how effectively this released uranium may travel will depend on many factors, not the least of which is climate, along with permeability of the rocks. For example, in the arid Southwest, there is a strong degree of oxidation, yet in many instances the uranium is not transported far because of lack of water. In the more humid great valleys of California, where water is abundant—either naturally or due to irrigation—much of the uranium has been flushed out of the soils. Thus, the uranium, and hence the radon, are both low. However, in the arid southwestern deserts, the uranium, the source for radon, is still in the rocks and sediment. There is another point to make here as well. In the desert case, the uranium is not only still close to its source, but commonly fixed in loosely bonded surface forms that can readily lose radon. This topic is discussed further in chapter 5.

Not all radon can be easily explained, however. Even if surface material is low in uranium, but is underlain by more uraniferous media, then a high radon flux at the surface is still possible. Similarly, in faulted terrain, the radon can migrate along the fault plane and cause local high fluxes. Still further, even deep-seated uranium can be detected by the presence of radon anomalies (see chapter 10).

Table 2.7 lists sources for radon in the United States, including natural and technologically enhanced sources, with uranium mining and milling listed separately. Also tabulated is the approximate dose to lungs for an average individual in the United States.

Two major sources of radon dose to lungs are obvious—natural soil and building interiors. Evapotranspiration (3 percent) and soil tillage and natural gas (each about 1 percent) constitute the rest, except for

TABLE 2.7
Sources of Radon in the United States

Source	Amount Released (Ci/y)	(PBq/y)	Approximate Dose to Lung from Radon and Daughters (%)
Natural			
Soil	(120,000,000)	(4,400)	40.0
Evapotranspiration	(8,800,000)	(330)	3.0
Technology Enhanced			
Building interiors	(28,000)	(1)	55.0
Soil tillage	(3,100,000)	(110)	1.0
Natural gas-industrial	(11,000)	(0.41)	1.0
Phosphate mining	(53,000)	(2.0)	0.02
Coal mines	(14,000)	(0.52)	0.005
Phosphate fertilizers	(48,000)	(1.8)	0.02
Geothermal Power	(580)	(0.021)	0.2
Uranium			
Mining	(200,000)	(7.4)	0.07
Milling	(150,000)	(5.6)	0.03

Source: Travis et al. 1979.

very minor amounts from phosphate mining, coal mining, phosphate fertilizers, geothermal power, and uranium mining and milling.

The data shown in table 2.7 may come as a surprise to some readers, but they actually make good sense. Soil releases 120 million curies to the atmosphere each year, and evapotranspiration adds another 8.8 million curies. The soil-released radon is more likely to be inhaled than that released from evapotranspiration; hence soil radon accounts for some 40 percent dose to lungs while evapotranspiration yields only 3 percent. On the one hand, indoor sources release only 28,000 Ci/year and account for some 55 percent of dose to lungs and include indoor radon as well. This reflects the fact that much of our lives is spent indoors. Tilling of soil, on the other hand, only adds one percent of the dose to lungs. Natural gas adds only 11,000 curies per year to the atmosphere, but, since it is used primarily for heating inside buildings, this adds another one percent to the lungs. The amounts of radon added from phosphate mining and fertilizers, from coal mines, and from geothermal power are small, and so are the doses from these sources to the lungs. Interestingly, prior to 1980, uranium mining and milling together added some 350,000 curies of radon per year, yet only 0.1 percent of the total dose to lungs was added, mainly because uranium mines are isolated from population centers (many uranium operations have since been shut down).

Radon escapes into the atmosphere from several large sources: land, oceans, vegetation, and waters on land. It has been estimated that some 2.4 billion Ci/year are released from continental land surfaces and some 23 million Ci/year from ocean surfaces, or the oceans contribute about one percent of the total from the lands. This is due to the fact that radium, the immediate progenitor to radon, is present in surface soils in an amount of 900 pCi/kg, while only 0.05 pCi/kg radium occurs in ocean water. Further, it is estimated that crops and other vegetation and groundwaters may add an amount equal to 20 percent of the land release value, or about 480 million pCi/yr

Radon is highly soluble in water, especially cold water, which helps its transport in geologic media. When water is warmed, the solubility decreases and radon gas is released (in showers, washers, etc.) and contributes to indoor radon in some areas (see chapter 5). Groundwater may typically contain radon levels of 5,000 pCi/L or so, and values much in excess of this are not uncommon, as 100,000 pCi/L has been found in some well waters in Maine. It is estimated that 10,000 pCi/L is responsible for about 1 pCi/L of indoor radon.

Once in the atmosphere, radon is fairly well mixed in local areas, but not as efficiently mixed worldwide. Thus, over areas of uranium mines, the radon is somewhat concentrated. Over oceans, typical radon measurements may vary from 0.002 pCi/L to 0.07 pCi/L, while typical air measurements over lands vary from 0.l06 pCi/L to 0.l3 pCi/L. Measurements in any one area may vary widely (measurements in New York City range from 0.02 to .5 pCi/L, a factor of 25). Out doors, people are exposed to somewhere in the range of 0.1 to 0.15 pCi/L.

The radon concentrations in the atmosphere vary. They are higher over continents than over oceans because of the higher radium content of soils and rocks compared to waters. Table 2.8 shows some average

TABLE 2.8
Some Typical Outdoor Radon Values (pCi/L)

Location	Value (pCi/L)
France	0.25
London	0.009
Germany (FRG)	0.07
Soviet Union	0.17
New Mexico	0.24
New York (city)	0.13 (0.02–0.50)
New York (rural)	0.21
Ohio	0.48
Washington, D.C.	0.12
Japan	0.06
Peru	0.04
Bolivia	0.04
Oceans	
North Atlantic	0.006
South Pacific	0.002
Indian Ocean	0.002
South Pole	0.001
Average North American Continent	0.2
Average Northern Hemisphere	0.1
Average Southern Hemisphere	0.1

Sources: NCRP 1975; Cohen 1979; U.S. Congress 1982; Sheearer and Sill 1969.

values for atmospheric radon. It is interesting to note that these data, limited as they are, do not correlate well with regional geology. For example, the value of 0.48 pCi/L for Ohio occurs where uranium favorability is low. When radon is released from soils, it is rapidly mixed vertically during the daylight hours and about one-third as efficiently mixed at night. Thus, radon levels near the earth's surface are higher at night than during the day. Radon will also be affected by winds, and these cause radon levels to be higher during the seasons with strong winds (usually spring and fall, with summer and winter somewhat lower). In coastal areas, when ocean winds predominate, the radon levels are low; when land winds occur, the radon levels are higher. Radon concentrations decrease exponentially with altitude, and values of 0.004 to 0.04 pCi/L are noted at elevations of 25,000 feet over the southwestern United States (compare to 0.24 pCi/L outdoors air average radon for New Mexico; see table 2.7). For this reason, radon is a useful tracer for measuring atmospheric mixing.

RADON EMANATION FROM SOILS: OVERVIEW

The manner in which radon is released from soils into the earth's atmosphere is understood in some generic sense, but this is an extremely complex problem. While this topic is covered in depth in chapter 5, I will give a brief overview here.

The source of the radon, i.e., parent uranium and/or intermediate daughter radium (^{226}Ra), is important. High uranium content may not correlate with high radon release; for example, the uranium may be housed in resistate minerals in the soil such as zircon, monazite, apatite, sphene, and others. In these minerals, the bonding is so strong that diffusion of radon from the original uranium site is vanishingly small. In the absence of any other radon source, the radon released from such a hypothetical situation may actually be inversely proportional to uranium content.

Yet, in many instances where weathering has caused rock-forming minerals housing some uranium to chemically alter, the uranium is oxidized and removed from the original mineral. The extent of migration of this uranium will be a function of soil permeability, moisture content, and other factors. In arid climates, the uranium is commonly fixed close to its source material; in more humid climates, the uranium may be transported considerable distances. This released uranium is often poorly collected on organic residual matter or on other particulate matter, at times forming surface films. Intermediate daughter loss from the ^{238}U chain is common here. Radium especially is easily removed from the ^{238}U chain. This radium then becomes a new, more convenient source for its immediate decay daughter, ^{222}Rn. In such cases the radon and radium will be in equilibrium, but the radon and/or radium may not be in equilibrium with uranium.

Regardless, when radon is in a site that is favorable for loss, this is not a guarantee that it will instantaneously escape from the soil. Soil moisture plays an important part here. Radon is released more readily from relatively dry soils than from wet soils, all other parameters being equal. This is due to the fact that the radon released from solids is captured in the water in which it is dissolved and transported.

Similarly, colder temperatures tend to retard radon migration, and thus cold soils release proportionately less soil than warm soils. Yet this factor is not large. The indoor radon problem in winter, for example, is much greater than in summer. This is a reflection mainly of how buildings are constructed, the living habits of the occupants, and other factors. But if the radon release from cold soils were very sluggish, it could have an effect on radon emanation during winter, although this has not been observed as a major factor.

It is also important to know if the soil is PERMEABLE. In highly permeable soil, radon is easily released. In relatively impermeable media, the opposite is true. In places where a house is built over alternating layers of shale and sandstone and where the shale is more uraniferous than the sandstone, much of the radon release will nevertheless be from the more permeable sandstone.

In the surface environment, the role of barometric pressure is important. It must be pointed out that soil radon gas is concentrated much more than in the atmosphere. Typical atmospheric radon values may be 0.1 pCi/L, but soil radon gas (at an assumed depth of 15 inches [38 cm]) is usually about 100 pCi/L. Thus there is a difference of 100 between the soil gas and the atmosphere. Consequently, when the pressure is high, atmospheric air tends to be driven into the ground. Under low pressure conditions, the opposite is true, and soil gas, including radon, is more readily released into the atmosphere.

Diurnal variations of radon emissions from soils are suspected, but these are only of minor importance.

Wind is somewhat an unknown factor as far as its effect on radon emanation from soil is concerned. Some data is available from studies of radon emissions from uranium mill tailings. In open, relatively low, relief areas, radon formed over the tailings is rapidly dispersed to background values within a kilometer of the pile. In areas of rugged topography, evidence for tailings-generated radon may be determined downwind for up to several kilometers. Thus, the simple observation to be made here is that if a source of high radon emanation is available, then downwind from it, the "average" radon level may increase somewhat; but for most land areas, this effect is probably small. A discussion of analogs to wind, i.e., fans and other ventilation systems for remedial cleanup of high indoor radon, is given in chapter 8.

RADIUM BUDGET

The radium budget (i.e., radon distribution and amount) in common rocks is given in table 2.9. There it is noted that granitic rocks (granite and rhyolite) have much higher radium contents than more basic rocks such as basalts, and alkalic rocks, such as syenites and phonolites, which have the highest values for igneous rocks. It is important to note, however, that shales (1973 pCi/kg) and most other sedimentary and metasedimentary rocks are also over 1000 pCi/kg. Thus, the radium contents of these rocks could lead to high radon release under appropriate conditions.

The average radon content of soils for the United States is given in table 2.10. The highest values are found in New Mexico (see chapter 7 for discussion of the Albuquerque area), Nevada, Ohio, and Kentucky. Colorado also has high values. The ranges are considerable for all the values given in table 2.10, however.

NATIONAL PROGRAMS

The National Uranium Resource Evaluation Project. The National Uranium Resource Evaluation project, or NURE, was conceived in the mid-1970s by the U.S. Atomic Energy Agency, and was carried over through the U.S. Energy and Resource Development Administration

TABLE 2.9

Radium Concentrations in Some Rocks

Rock	No. Samples	^{226}Ra (pCi/kg) Arithmetic Mean
Rhyolite	131	1919
Granite	569	2108
Andesite	71	702
Diorite	271	1081
Basalt	77	297
Gabbro	119	270
Dunite	31	108
Phonolite	138	9946
Syenite	75	18702
Nephelinite	27	784
Foidite	8	784
Evaporites	243	1216
Limestone	141	676
Clay	40	1351
Shale	174	1973
Sandstone and conglomerate	198	1378
Gneiss	138	1351
Schist	207	1000

Source: Wollenberg 1984.

(ERDA) and the U.S. Department of Energy (DOE). Among other purposes, a major goal was to sample the entire contiguous United States and Alaska with a water and/or stream or other sediment sample per 10 square kilometers for this entire area. Each sample was analyzed for uranium and thorium, and many were analyzed for additional trace elements as well. The purpose was to discover areas of high background uranium, which might, in turn, indicate areas of potentially favorable uranium deposits. The results of these studies have been published by the four federal laboratories responsible for the sampling and analyses: Savannah River Laboratory, Oak Ridge National Laboratory, Los Alamos National Laboratory, and Lawrence Livermore National Laboratory.

While the use of NURE to the discovery of economic concentrations of uranium can be argued, this project nevertheless has provided a very

TABLE 2.10

^{226}Ra Concentrations in Soils in the United States

Rock	No. Samples	^{226}Ra (pCi/kg) Arithmetic Mean
Alabama	8	811
Alaska	6	649
Arizona	6	946
Arkansas	0	
California	3	757
Colorado	32	1405
Delaware	2	1162
Florida	11	838
Georgia	9	892
Idaho	12	1108
Illinois	7	973
Indiana	2	1054
Kansas	6	973
Kentucky	13	1514
Louisiana	2	703
Maryland	6	730
Michigan	10	1108
Mississippi	3	1189
Missouri	10	1108
Nevada	6	1514
New Jersey	24	865
New Mexico	113	1514
New York	6	838
North Carolina	8	784
Ohio	12	1514
Oregon	8	811
Pennsylvania	33	1189
Tennessee	10	1108
Texas	10	892
Utah	32	1297
Virginia	13	838
West Virginia	11	1297
Wyoming	13	1000
All samples	327	1108

Source: Myrick et al,. 1983.

useful data base for background uranium in the United States. In the arid western and southwestern states in particular, the sediments sampled far outweigh the number of waters, and, further, the soils in many areas are developed in immature fashion from sediment. Hence, the sediment uranium values are equal or close to soil uranium values, which, in turn, means that areas of high uranium can be identified, and then these areas can be tested for potentially high radon emanation. Figures 2.2 and 2.3 are maps from the DOE which show favorable geology for uranium occurrences in the United States as well as the potentially promising areas. Note that while most of the promising uranium areas are in the western United States, the conditions for favorable geology occur over much of the country. Indeed, figure 2.3 does not show the uranium occurrences of western Nebraska nor the vein type deposits discovered in Virginia. The rocks of the eastern United States can, in many instances, be highly uraniferous. The Reading Prong has

FIGURE 2.2
Map of the United States showing areas (dark) favorable for uranium occurrences of economic potential. These areas may be prone to radon emissions. Modified from DOE.

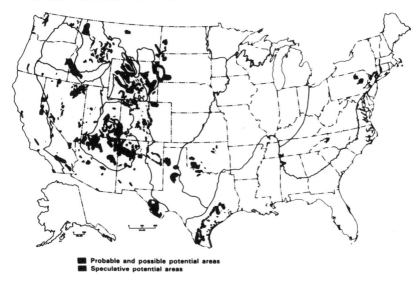

■ Probable and possible potential areas
■ Speculative potential areas

NATIONAL URANIUM RESOURCE EVALUATION - POTENTIAL URANIUM AREAS

FIGURE 2.3
Map of the United States showing those areas (dark) with favorable geology for possible uranium occurrences. These areas may be prone for radon emissions. Modified from DOE.

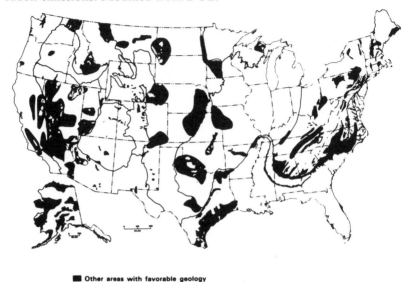

■ Other areas with favorable geology

NATIONAL URANIUM RESOURCE EVALUATION - AREAS WITH FAVORABLE GEOLOGY

already been mentioned as one such rock. The Conway and other granites of New Hampshire and elsewhere in the Appalachians are others, as is the Chattanooga Shale of the mideastern United States and the phosphate bearing rocks of Florida, Georgia, and Tennessee. Wherever such rocks occur, the potential for radon release may be even higher. An important factor here, of course, is how the uranium, and thus the radon, is bonded in the rocks. If tightly bonded, then the radon release potential might be low—possibly lower than from lower uranium-bearing rocks, but with higher release potential for radon. These ideas are discussed elsewhere in this chapter.

Unfortunately, a detailed map of the United States showing the background uranium values has not been put together. Despite the formidable nature of such a job, it could prove to be a most useful exercise, especially if it assists the EPA in pointing to areas where radon potential is high.

One final note on NURE. I, along with many others, have written to various agencies in Washington in an attempt to get HEW, NIH, and so on, to participate in the program so that multi-elements could be analyzed for all the samples collected. This would give the United States an incredible data base for assessing background geochemistry for topical problems of rock geochemistry and patterns in disease and deaths and allowing assessment of future contamination of areas (i.e., at present it is often difficult to prove contamination because background for so many elements is poorly known). This has not been done. But the government might be able to use this extensive data base for radon-related work.

The National Airborne Radiometric Reconnaissance Program. The NURE project had several subparts. One of these was the National Airborne Radiometric Reconnaissance (NARR) program, which was designed to identify, by airborne radiation detecting devices, areas of high naturally occurring radioactive materials. This information would then be used to help identify areas for follow-up uranium investigations.

The NARR data provide a useful, although somewhat limited, approach to identifying areas of high soil radium, the immediate precursor to radon. The researchers (see readings for this chapter in appendix) found that in over 50 percent of the cases, the NARR data identified areas of high radium that could be correlated with high radon flux, but that this was not always the case. In cases where the radium has been added to soil recently, for example, its long half-life of 1625 years means that not enough radon has formed to correlate with the radium. In older soils, however, the radium is useful because it has reached some kind of equilibrium with radon. When people till soil or turn over large amounts of land for building purposes, much of the radon is lost (as described earlier in this chapter). Hence, although time-consuming, obtaining uranium values may be of more use than obtaining radium values.

Nevertheless, the NARR data have been useful. Workers have used the data to identify areas of high potential radon in parts of California, Oregon, Washington, Idaho, and Nevada. They have had about 50 percent success. In Virginia and Maryland, the U.S. Geological Survey's airborne radiometric maps are useful in identifying areas of high radon potential, although it has been pointed out by Dr. Douglas Mose, of George Mason University (written communication), that, among others, follow-up soil studies are needed to clearly identify specific areas in part because lithologic changes in rocks are the rule rather than the exception, and thus uranium contents can vary widely in a single unit of rock. Similarly, groundwater hydrology can play a major role, and this factor becomes a variable within other areas of comparable geology.

CONCLUSIONS

Uranium is a trace element in the earth's crust, but it is very widely distributed. All rocks contain some uranium, and uranium is the ultimate parent for radon. Hence all rocks, and all soils, contain some potential for radon release.

Different rocks, however, contain different amounts of uranium, and often the uranium-bearing minerals are sporadically distributed. Yet, by recognizing certain rocks as prime candidates for radon release—granite for example—areas of high potential radon emanation can be identified. When uranium has been lost from a rock, usually due to combined oxidation and groundwater transport, the resulting rock may have a high thorium to uranium ratio indicating this loss. Care must be taken so that it is also ascertained that the radium, formed from uranium and parent radon, is also absent from the rock. By combining chemistry and geology, it is possible to gain insight into the radon potential for different areas.

Radon in waters is also variable, but in general, the radon from this source is of secondary importance relative to radon emitted from soil. How radon is emitted from soil is complex (see chapter 5), but ways to identify high radon-emanation soils are being found.

The National Uranium Resource Evaluation (NURE) program of the DOE has provided a huge data base for sediments and surface (or shallow) waters for uranium in the contiguous United States. Use of these data should allow EPA and other government agencies to zero in on areas for radon testing, although this approach has so far not been carried out. I have examined the NURE data for parts of New Mexico, for example, and found a positive correlation of soil radon and indoor radon with uranium content.

While the distribution of uranium is complex, it governs just where radon will be released. Hence, an understanding of the uranium budget of the earth is fundamental to the problem of indoor radon. Fortunately, we do have a great deal of information about uranium in the United States, and radon data are beginning to be rapidly obtained as well.

3

Radioactive Decay of Uranium and Thorium; Radon and Radon Daughters

The three important radon isotopes, ^{222}Rn (the most important), ^{220}Rn (thoron), and ^{219}Rn (actinon), are intermediate radioactive daughter products of ^{238}U, ^{232}Th, and ^{235}U, respectively. In this brief chapter, I will discuss the total radioactive decay chain of ^{238}U only (see fig. 3.1), although the decay chains for ^{232}Th and ^{235}U (figs. 3.2 and 3.3) will be commented on as appropriate.

RADON AND RADON ISOTOPES

Radon, an inert gas under normal chemical conditions (see introduction), contains twenty-six known isotopes, all of which are radioactive. In nature, ^{222}Rn is the most abundant and the most important. Both ^{220}Rn, sometimes called thoron, and ^{219}Rn, sometimes called actinon, are much less abundant and of much less health significance than ^{222}Rn. The half-life of ^{222}Rn is 3.8 days, which, while short compared to long-lived radioactive isotopes such as ^{238}U (half-life of 4.5 billion years) or ^{226}Ra (half life of 1,625 years), is nevertheless much greater than the half-lives for ^{220}Rn (55.6 seconds) and ^{219}Rn (3.96 seconds). On the one

FIGURE 3.1

Decay scheme for uranium 238.

Mass	Elements/Isotopes (half lives)
238	Uranium
	4.468×10^9 y
234	Thorium → Protoactinum → Uranium
	24.1 d 6.7 h 2.446×10^5 y
230	Thorium
	7.54×10^4 y
226	Radium
	1625 y
222	Radon
	3.82 d
218	Polonium
	3.05 m
214	Lead → Bismuth → Polonium
	26.8 m 19.9 m 164 us
210	Lead → Bismuth → Poloni[...]
	alpha decay 22.3 y 5.013 d 138.4
206	beta decay → Lead
	stable

Modified from Brookins (1984) and Seelmann-Eggebert et al., (1981)

Abbreviations: y = years, m = months, d = days, h = hours − seconds, us = microseconds

hand, in a typical soil at some modest depth, the radon gas form of ^{220}Rn or ^{219}Rn decays to its radon daughters so quickly that only small amounts reach the atmosphere. Radon-222, on the other hand, readily escapes into the atmosphere.

In certain thorium-rich areas of the world, ^{220}Rn may also be of concern for health (see references for this chapter). In chapter 2 I noted that uranium is about four times as abundant as thorium, yet the short half-life of ^{220}Rn prevents its ready escape into and accumulation in the earth's atmosphere.

The amount of ^{235}U in the earth is small, making up only 0.7 weight percent of total uranium. Thus, this factor, combined with the very short half-life of ^{219}Rn, makes it essentially of no significance for public health concerns.

FIGURE 3.2

Decay scheme for uranium 235.

Mass	Elements/Isotopes (half lives)
235	Uranium
	7.038×10^8 y
231	Thorium → Protoactinum
	25.5 h 3.276×10^4 y
227	Actinium → Thorium
	21.77 y 18.72 d
223	Radium
	11.43 d
219	Radon
	3.96 s
215	Polonium
	1.78 ms
211	Lead → Bismuth
	36.1 m 2.17 m
207	Thalium → Lead
	alpha decay 4.77 m *stable*
206	beta decay →

Modified from Brookins (1984) and Seelmann-Eggebert et al., (1981)
Abbreviations: y = years, m = months, d = days, h = hours − seconds, ms = milliseconds

Support for considering only ^{222}Rn as a major health concern can be found in the recent study of the NAS (1988), in which it is noted that given a dose to the lungs for equal alpha energies of ^{222}Rn and ^{220}Rn, the risk from ^{222}Rn is three times that from ^{220}Rn. Therefore, in nature, where the amount of energy from ^{222}Rn is substantially greater than that from ^{220}Rn, the thoron contributes so little relative damage compared to that from ^{222}Rn as to be negligible.

RADIOACTIVE PARTICLES AND RAYS

Three types of radioactive decay are possible: ALPHA, BETA, and GAMMA. An alpha particle is essentially a helium nucleus $^4_2$He^{2+} with a charge of $+2$ and a mass of 4. It is given off when very heavy metals

FIGURE 3.3
Decay scheme for thorium 232.

Mass	Elements/Isotopes (half lives)
232	Thorium
	1.405×10^{10} y
228	Radium → Actinum → Thorium
	1.913 y
224	Radium
	3.66 d
220	Radon
	55.6 s
216	Polonium
	0.15 s
212	Lead → Bismuth → Polonium
	10.64 h 60.6 m 14.2 ns
208	Thalium → Lead
	alpha decay 3.053 m Stable
206	beta decay

Modified from Brookins (1984) and Seelmann-Eggebert et al., (1981)
Abbreviations: y = years, m = months, d = days, h = hours – seconds,
ns = nanoseconds

undergo radioactive decay. Thus, in the decay of ^{238}U to the (eventual) stable lead isotope (^{206}Pb), since each alpha particle has a mass of 4, 8 alpha particles are given off. On the one hand, as shown in figure 3.1, every time an alpha particle is emitted, the parent (including intermediate radioactive daughters) changes in mass by 4 and in atomic number by 2. This is because two protons and two neutrons are emitted in the alpha particle. A beta particle, on the other hand, is essentially an electron. Its mass is only 1/7500 of the alpha particle, and it is obviously much smaller. It is for this reason that the beta particles can readily pass through more matter than alpha particles can. For example, it will take three feet or so of concrete to stop a beta particle; whereas, an Alpha particles can be stopped by a sheet of paper. Gamma radiation is a form of x-radiation of extremely small mass. Therefore, gamma rays can penetrate even further than beta particles, but the damage imparted to the median through which they pass is minor.

As the alpha particles are emitted, they travel at very high speeds and come into contact with matter. As this interaction proceeds, electrons are knocked off and ionization results. In air, alpha particles move about 2.5 centimeters, but in the more dense tissue, only on the order of tens of microns. It must be pointed out, however, that not in all cases is air radon in equilibrium with air particulates to which radon daughters are attached. Thus, if some particulates are removed (regardless of process) but a continuous source of radon is provided, then, assuming equilibrium, amounts of radon daughters per amount of radon present can be in error.

Beta and gamma radiation penetrates matter to a much greater extent than alpha particles, but their effect on tissue is proportionately much less. Beta particles are electrons or, more rarely, positrons, with very small mass compared to the alpha particle.

While bismuth (Bi) and lead (Pb) radioactive isotopes produced by the decay of radon are not alpha-emitting, they are important because they, like Po isotopes, have short half-lives such that they decay before lung clearing mechanisms can effectively remove them. This is discussed in some detail in chapter 4.

In terms of damage to matter subjected to alpha, beta, and gamma radiation, the alpha particle causes the most damage. Imagine an elephant and a field mouse entering a field of wheat. The field mouse, an analog for the beta particle, does little damage to the wheat; the elephant, an analog for the alpha particle, does major damage. If we use the light from a flashlight as an analog for the gamma rays, then the light may be seen through a lot of wheat, but essentially no damage to the wheat results from the passage of the light. In a very approximate way, my wheat field scenario demonstrates what happens when alpha particles of high energy strike human tissue. The damage can be considerable. In chapter 4, I will address the health aspects in detail. Here, suffice it to say that the alpha-emitting isotopes taken into one's lungs can, with high levels of such alpha particles over a long period of time, result in major damage to the lung tissue—and some 15,000 individuals in the United States each year die of lung cancer from this cause.

The penetrating power of alpha, beta, and gamma radiation is illustrated in figure 3.4.

RADON PARENTS

As shown in figure 3.1, ^{238}U decays by a series of complex alpha and beta emitting steps to eventually form ^{206}Pb. The formation of ^{222}Rn occurs about half-way through this complex chain. Before it is formed, however, numerous other isotopes of different elements are formed. In figure 3.1, the half-lives of the isotopes in the ^{238}U chain are shown. The ^{238}U has a very long half-life, 4.5 billion years, so the rate of its decay is very slow. Interestingly enough, the age of the earth is also about 4.5 billion years, so uranium has had enough time to form copious amounts of all the isotopes shown in figure 3.1. The ^{238}U first decays by alpha emission to thorium- 234 (^{234}Th) and then to protoactinium-234 (^{234}Pa) by beta emission and thence by another beta emission to ^{234}U. This is an important part of the whole chain, for the ^{234}U formed is more soluble than the ^{238}U due to a change in valence from $+4$ to $+6$ (see chapter 2). Thus, if the uranium is on the edge of a grain, where water can contact it, it may be transported away in groundwaters or surface waters. By alpha decays, then, ^{230}Th is formed (half-life of 80,000 years), and then ^{226}Ra with a half-life of 1625 years. Both thorium and radium are very insoluble in the earth and can be fixed on grain surfaces, cracks, and other areas where the soluble ^{234}U may have migrated. Thus, when the decay of ^{226}Ra to ^{222}Rn takes place, the radon formed may be in a site where it may readily escape into pore spaces as a gas or be dissolved in water and transported and later released. This is how most of the radon that escapes to the earth's atmosphere is formed (chapter 5).

The importance of the radon parents, then, is that their behavior in rocks, minerals, and soils controls to a large degree the favorable siting conditions for radon release. Also, once the immediate parents, ^{230}Th and ^{226}Ra, are fixed, because they are highly insoluble and have very long half-lives, a built-in supply of radon is assured for many thousands of years.

RADON DAUGHTERS

Unlike several of the immediate parents to ^{222}Rn that have very long half-lives (^{238}U, ^{234}U, ^{230}Th, ^{226}Ra), the daughters of ^{222}Rn have short half-lives. This is shown in figure 3.1. By alpha particle emission, ^{222}Rn decays to polonium-218 (^{218}Po) with a half-life of 3.05 minutes; then this goes by alpha decay to lead-214 (^{214}Pb) with a 26.8 minute half-life, which undergoes beta emission to form bismuth-214 (^{214}Bi); this, in turn, undergoes another beta emission to form polonium-214 (^{214}Po), which undergoes alpha emission to form 22.3–year lead-210 (^{210}Pb). This last isotope, then, by two beta emissions, goes to ^{210}Bi and ^{210}Po. Finally, emission of an alpha particle from ^{210}Po forms stable lead-206 (^{206}Pb).

Of these radon daughters, the two that may cause the most damage to tissue are the isotopes of polonium, especially ^{218}Po and ^{214}Po. This is because both are alpha-emitting isotopes, and, again, alpha particles can do great damage to tissue, whereas under most conditions, the damage imparted by beta particles is often so minor as to be almost negligible. Table 3.1 indicates the decay energies of the radon daughters along with their respective half-lives. The four alpha decays, from ^{222}Rn, ^{218}Po, ^{214}Po, and ^{210}Po, are much greater than the decay energies for the beta-emitting isotopes. Polonium-218, in addition, is formed as a charged ion, ^{218}Po^{2+}, which means that it can more easily be affixed on particulate matter and can react with that matter—such as lung tissue. The decay energies are given in million electron volts (MeV), which is one

TABLE 3.1
Radon and Radon Daughter Isotopes from ^{238}U Decay

Isotope	Decay Process	Decay Energy (MeV)	Half-life
^{222}Rn	alpha	5.49	3.82 d
^{218}Po	alpha	6.00	3.11 m
^{214}Pb	beta	0.7 (est.)	26.8 m
^{214}Bi	beta	2 (est.)	19.9 m
^{214}Po	alpha	7.69	164 us
^{210}Pb	beta	0.04 (est.)	22.3 y
^{210}Bi	beta	1.16	5.01 d
^{210}Po	alpha	5.30	138 d
^{206}Pb	stable	—	—

Sources: Seelmann-Eggebert et al. 1981; Nero 1987; and others.
Abbreviations:
  d = day
  m = minute
  us = microsecond
  y = year

FIGURE 3.4
Penetrating power of alpha, beta and gamma radiation. Alpha particles are stopped by a sheet of paper, beta particles by a thin sheet of aluminum, and gamma rays by three feet of cement. See text for details.

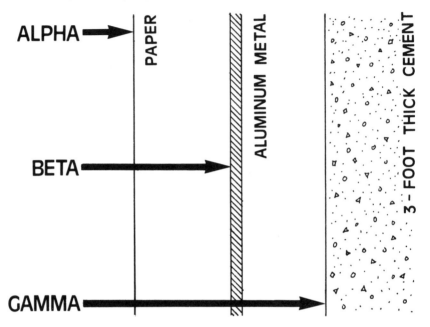

conventional way of reporting such energies. The importance here, however, is the relative energies of the alpha-emitting versus the beta-emitting isotopes.

When we inhale air, we take in a certain amount of radon and, proportionately, an amount of radon daughters. While we exhale or trap out most of these (see chapter 4), we continue to breath, and the amount of radon and radon daughters that reaches our lungs is, ultimately, proportional to the concentration of these isotopes in the air around us. Again, this is the crux of the indoor radon problem: High levels of radon mean high levels of radon daughters and a probability of high amounts of these reaching and potentially damaging our lungs.

CONCLUSIONS

While radon is formed from the decay chains of ^{238}U, ^{235}U, and ^{232}Th, only ^{222}Rn, the daughter of ^{238}U, is important to maintaining good health. As pointed out in chapter 2, ^{238}U is 138 times as abundant as ^{235}U, and the ^{219}Rn formed from ^{235}U has an extremely short half-life of 3.96 seconds. So, too, does ^{220}Rn, the daughter of ^{232}Th, with a 55.6 second half-life. Relative to ^{222}Rn, then, the ^{219}Rn and ^{220}Rn largely decay to their daughters prior to escapt to the atmosphere.

4

Health Effects of Radon and Radon Progeny

In this chapter, I will briefly explore the health effects of radon and radon daughters in promoting lung cancer. We have already seen (chapter 3) that alpha particles, such as those given off by ^{222}Rn, ^{218}Po, and ^{214}Po, can cause considerable ionization in matter compared to the less energetic beta particles and much less energetic gamma rays. This ionization can be thought of as damage to the medium through which the alpha particle moves. In the human lung, some alpha particles move through the lining membrane in the bronchial region and, if abundant alpha particles are involved, may cause cell abnormalities that may become tumerous.

A large number of lung cancer fatalities occurs in the United States and in the world (see below), but not all these lung cancers are due to smoking. While smoking is certainly the major culprit, an appreciable number of lung cancers may be caused by radon and radon daughters. A smaller but no less significant number, of lung cancer deaths may be caused by other carcinogens such as metal, some organic compounds, and a very few by blue asbestos (crocidolite).

The risk of radon-related cancer to the general public is based on studies of uranium miners and other miners exposed to radon in high amounts over long periods of time. Further data must be carefully weighed to correct for the miners who smoked and other factors. It may be possible to detect lung cancers caused from radon versus lung cancers caused from smoking, but such studies are not yet complete.

As this chapter will show, lung cancers from indoor radon are significant. In fact, they are of large-scale proportions, and nationwide education is needed to help reduce the number of lung cancers exacerbated by indoor radon.

LUNG CANCER IN THE UNITED STATES

There are about 200,000 lung cancer fatalities in the United States each year. The rate of lung cancers per 100,000 population for men and women is shown in figure 4.1. Men are not more susceptible to lung cancer than women; differences in the two curves reflect that, historically, men smoked much more than women. It is interesting that not only is the curve for women on a marked increase, but the curve for men has started to taper off. Approximately 150,000 of the total lung cancers per year are thought to be due to smoking.

The National Academy of Science (NAS) estimates that in the United States at least 15,000 fatal lung cancers per year are due to indoor radon, and another estimate is that 5,000 lung cancer fatalities occur from passive smoking. These are all large and frightening figures.

Another viewpoint on lung cancers, according to the EPA, is that asbestos is a contributory factor. While asbestos is not the focus of this book, I must point out that in this matter the EPA is misinformed. Malcolm Ross (1984) of the U.S. Geological Survey has addressed this problem in detail, and his data, now supported by studies from England, show that the common white asbestos, chrysotile, mined and used widely in North America, does not contribute significantly to lung cancer. Only blue asbestos, or crocidolite, from South Africa and Australia, poses a real threat. Ross convincingly shows that the actual lung cancers from all varieties of asbestos in the United States each year is well under 100 —which is a very small figure when compared to the 15,000 radon-caused and 150,000 smoking-caused illnesses.

THE LUNGS

Figure 4.2 shows a brief sketch of how air is taken into the human body. We inhale air through the nose or mouth, and it passes downward

FIGURE 4.1
Age adjusted lung cancer death rates for males and females in the United States 1930–1984. Data from U.S. Bureau of the Census. Modified from DOE.

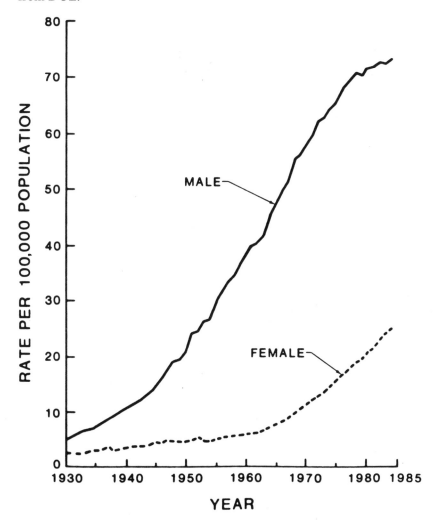

through the pharynx and trachea. The air then divides at the entrance to the bronchial system; each of these two pathways continues to divide into branches millions of times and eventually end in very tiny air sacks or pockets called aveoli. The final bronchial passages that end in the

FIGURE 4.2

Sketch of the lungs showing air pathways through trachea, the larynx, bronchi, bronchioles, and aveoli. See text for details.

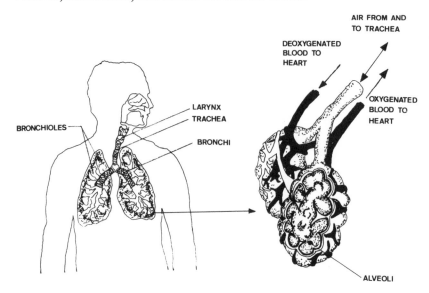

aveoli are measured in microns (i.e., one micron $= 10^{-6}$ meters) because they are so small.

Close to the mouth, the first few divisions of the bronchi are lined with cilia—tiny hairs that filter out most of the airborne particulate matter. In fact the cilia are very effective in removing particulates and

FIGURE 4.3

Sketch of the bronchial epithelium showing its layered structure, and a protective coating of mucous. For comparison, the bars for the penetrating range of some radon and radon daughter alpha particles are also shown.

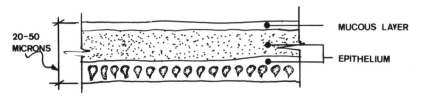

60-80 MICRONS: PENETRATING POWER OF PO-ALPHA PARTICLES

actually cause much of the particulate matter to be exhaled within hours of inhalation. Earlier, considerable dust and absorbed smaller particles are removed by nose hair. In the case of radon daughters, however, these decay with half-lives of minutes, seconds, and fractions of seconds, thus emitting their radiation before they can be cleared. It is no accident that much of the lung cancer in miners occurs in the upper bronchial passages where the cilia have fixed the radon daughters.

The membrane that lines the tissue in the lungs is called the epithelium and is made up of two layers: a basal layer, usually about seven microns thick, and an outer layer that is 40 to 80 microns thick. The outer epithelium is covered by a mucus layer of variable thickness (estimates are from 10 to 20 microns) (see fig. 4.3). The outer layer is very important because it can prevent some material from reaching the basal epithelium. In the outer layer, cell division (mitosis) does not occur frequently, but in the basal layer it is pronounced. Hence, in the outer layer, the carcinogenesis, or development of cancer cells, is much less probable than in the basal layer. In addition, the outer epithelium is often covered by a mucus layer which ranges from 10 to 20 microns thickness.

The mucus layer and outer epithelium are of extreme importance because they effectively armor the basal cells by their combined average thickness of about 60 microns. When the combined mucus and outer epithelium layers are near 20 to 30 microns, the alpha particles can reach the basal cells; when the layers are over 70 microns, then the basal cells are protected.

The importance of the bronchial epithelium is shown in figure 4.4. While the radon daughters account for 55 percent of the overall dose to the lungs, they account for 97 percent of the dose to the bronchial epithelium.

For the lungs, the three alpha-emitting isotopes of concern are ^{222}Rn and the radon daughters ^{218}Po and ^{214}Po. Their alpha energies and range in tissue are given in table 4.1, where it is noted that all three can easily penetrate to at least 41 microns, while ^{214}Po can penetrate to a depth of 78 microns. Although not covered here, it is interesting to note that the radon daughters of thoron (^{220}Rn) also penetrate tissue to depths of 50 to 90 microns. Since the bronchial epithelium is some 40 to 80 microns thick, all these isotopes can penetrate some of the thickness, and some can penetrate all of it.

Just where the radon and radon daughters may become affixed to the

FIGURE 4.4

Percentages of the annual dose to the public from exposure to natural background ionizing radiation. Source: DOE.

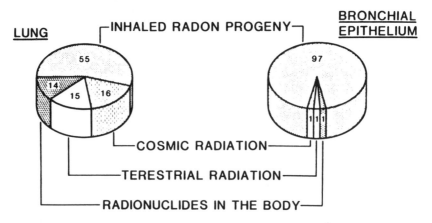

bronchial epithelium is also of concern. In the outer bronchial passages, the epithelium is about 80 microns thick, but this gradually narrows until it is only 15 microns or so thick in the deep lung near the aveoli. Thus, if radon or radon daughters manage to penetrate into the deep lung, then all alpha-emitting isotopes have a good chance of penetrating the epithelium and becoming available to attack cells.

It is not known precisely how alpha particles cause lung cancer. It is postulated that, due to their large mass, as these alpha particles enter the cell-rich area, they must encounter chains of the functional protein building material known as deoxyribonucleic acid (DNA), which consists of entwined chains of molecules. As shown in figure 4.5, whereas a beta particle might pass through a DNA chain without contact or, at

TABLE 4.1

Alpha Particle Ranges from Radon Daughters

Isotope	Energy (MeV)	Range (microns)
^{222}Rn	5.49	41
^{218}Po	6.00	48
^{214}Po	7.69	71

Source: Nazaroff and Nero 1987.

FIGURE 4.5

Interactions of beta and alpha particles with DNA chains. The beta particle either misses or causes insignificant damage to the DNA chain whereas the much larger alpha particle causes multiple breaks in it. See text for details. Modified from Weiffenbach (1982).

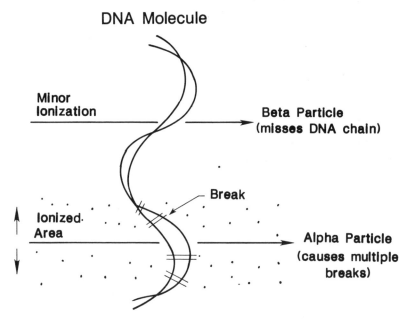

DNA Molecule

worst, merely impart a slight damage to the chain, the alpha particle will break the DNA chain in several places. When this happens, nature tries to put the severely damaged chain back together, but doesn't do it correctly. The imperfectly reformed DNA chain can then give rise to abnormal or tumorous cells. A single break in a DNA chain can usually be self-repaired without any permanent cell damage. Not so for the four or so breaks caused by the alpha particles.

A further complexity comes from the nature of the inhaled particulates. Charged particles are easily attracted to surfaces and fixed there; ^{218}Po (a charged isotope; see chapter 3) is thus readily attached to tissue. Uncharged particulates behave according to their size. It has been observed that only eight percent or so are deposited in the bronchial regions. Interestingly, in a dusty environment, such as mines or industrial work places, charged particles (such as ^{218}Po) are readily fixed to normal dust particulates prior to inhalation, yet in a low dust environment such as a house, the opposite may be true.

The dose of radiation received by the lung from outdoor sources is estimated at 50 mrem/year from ^{222}Rn and its daughters and perhaps as much as 5 mrem/year from ^{220}Rn and its daughters. The lung is the principal recipient of the radon and radon daughter radiation dose, although a very small amount of both is dissolved in the bloodstream and thus travels to other organs (at 1 to 3 order of magnitude lower doses). The National Council on Radiation Protection (1975) estimates that the radon gas and daughters dissolved in food and water add only 0.4 mrem/year to the average whole body.

RADON HEALTH EFFECTS:
STUDIES OF URANIUM MINERS AND OTHER MINERS

The estimates of low-level radiation, such as might be caused by the inhalation of radon gas and radon daughters, and its effect as a carcinogenic agent for attacking one's lungs are based on extrapolations from high-level radiation doses. Fatal lung diseases among miners have been known for centuries. Lundin, Wagoner, and Archer (1971) note that among miners in the Erz Mountains of Central Europe, as early as the mid-1500s, high incidence of lung diseases prevailed. They estimate radon daughter levels of 30 to 150 WL, or roughly 2000 to 30,000 pCi/L. In mines, especially before the twentieth century and even in the early part of this century, ventilation systems were poor at best, and presumably radon levels were extremely high in many hard rock mines. Unfortunately, even in post-World War II times, there is still a very high incidence of lung diseases in miners of the Erz Mountains.

Uranium mining on a significant scale started in the United States in the late 1940s, and radon levels were noted as being comparable in places to those in the Erz Mountains. Yet little was done to alleviate this potentially fatal situation until the 1960s, when the first signs of excess lung cancers started to show up. At about this time, the radiation safety limit of 1 WL was lowered to 1/3 WL.

The studies of uranium miners are very complex and at times difficult to explain. Figure 4.6 shows a well-repeated situation in which there is a linear dose to adverse health effect (i.e., lung cancer excess) relationship. This figure was based on the study of several thousand uranium miners

FIGURE 4.6

Risk of lung cancer for United States uranium miners, 1951–1971; modified from National Academy of Science (1988); see text for details.

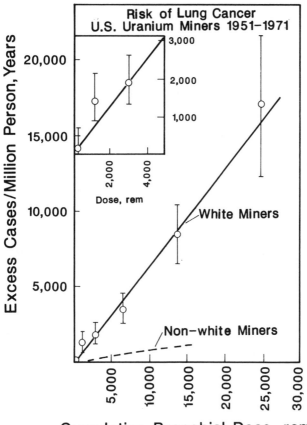

actively engaged in such mining during the 1950s and 1960s. Unfortunately, many of these miners had prior mining experience when they were almost certainly exposed to radon gas, but at what level and for how long is unknown. Also shown in figure 4.6 is a dashed curve for Native Americans, which is markedly lower.

In figure 4.6, the dashed curve for nonwhite (i.e., Native American) miners shows a much lower slope curve for lung cancer excess as a function of dose than the curve for white miners. This difference, at first thought to be something of a real genetic nature, turns out to be due actually to differences between smoking and nonsmoking. The Native Americans in general did not smoke, the white miners did.

Studies of Czechoslovakian miners are easier to understand. Of miners who started to mine uranium in the period 1948–1952, all were white and only one percent had prior mining experience. This group has been studied by several investigators who note a clear excess lung cancer factor proportionate to radiation dose.

Studies of fluorspar miners from Newfoundland, iron miners from several countries, Swedish metal miners, United States metal miners, and Canadian uranium miners (see summary in Cohen 1979) reveal a wide set of lung cancer data. A rate of lung cancer factor that is high not only for uranium miners but also for the fluorspar and metal miners has been calculated. In fact, the rates are higher than rates of lung cancers from the atomic weapons used in Japan at Hiroshima and Nagasaki and for the spondylitis patients treated with massive radiation doses in England in the late 1950s.

Miners are more often than not smokers, and usually heavy smokers. They are also exposed to fumes and dust, and incidences of silicosis and other lung affecting diseases are common. Further, many miners are exposed to carcinogenic chemicals (cadmium, lead, chromium, nickel, etc.) in the mine environment, and all of these nonradiation effects may be important. Choice of a control group is thus problematic, for, while the nonminers chosen are not exposed to the same underground radon levels, they are also not exposed to the fumes, dust, chemicals, etc. In short, the differences may be somewhat skewed to a greater difference than really exists.

For several years, Saccomanno et al. (1986) have investigated in detail the possible effects of cigarette smoking as augmenting the high incidence of lung cancer in former uranium miners. Their study tested over 9,000 individuals who had worked in underground uranium mines in the period 1960–1980. These individuals agreed to participate in the study and have been monitored continuously except when cases of death (all causes) or other extenuating circumstances, such as relocation, occurred. Their findings are of extreme interest. Adjusting for age, length of time worked underground, amount of cigarettes smoked, cumulative exposure to radon and radon daughters, and other factors, the authors note that the suspected causative effect of smoking on top of high radon exposure was not demonstrated. Further, the investigation showed that below 300 WLM exposure, the non-cigarette-smoking group had a lung cancer risk equivalent to the general, nonsmoking population. Thus, if a person did not smoke and worked in underground uranium mines for a period of time such that he or she received less than 300 WLM (i.e., at 1 WL per month, would require 25 years mining), the expectancy of that person getting lung cancer from radon in the uranium mines is equivalent to that of the general public who don't smoke. This substantiates the lower of the two curves for uranium miners (inadvertently broken down into "white" and "nonwhite) in figure 4.6.

The U.S. Department of Energy, however, using a mathematical model, notes that there is a pronounced increase of potential lung cancer risk with smoking. This is shown in figure 4.7. Here it is clear that for two miners, one who doesn't smoke and one who smokes three packs of cigarettes a day, there is a great difference in the risk involved. The heavy smoker's chances of getting lung cancer are 60 times as great as that of the nonsmoker, even though both are exposed to the same amount of radon and radon daughters over the same period of time. This seems to substantiate that there is indeed a positive correlation of excess lung cancer fatalities among miners due to combined radon and radon daughter exposure and smoking.

While this is somewhat predictable and data can be obtained from former miners, the smoking-radon correlation in the general public is not so clear. It is logical to argue that there is a multiplicative relationship between the two. Smoke is an excellent getter (accommodator) for charged particles such as radon daughters, and even radon can be affixed to smoke aerosols. Thus, as smoke enters and tries to coat the lung lining, radon and radon daughters are easily fixed as well. Of interest, though, is the observation that smoking often increases the mucus in ones lungs (so-called smoker's cough is a result), and this may actually help prevent some of the alpha particles from penetrating the bronchial epithelium. The multiplicative effects of smoking and indoor radon cannot be denied, however. The National Academy of Science (1988) suggests a strong correlation between the two. Certainly, the studies of uranium miners who smoked heavily show that this group had the largest incidence of lung cancers relative to all other groups. There is a long way to go before this smoking-radon relationship is resolved, but in the absence of other criteria, it makes good sense not to smoke

FIGURE 4.7
Predicted ratio of lung cancer risk for miners who smoke cigarettes and inhale radon compared to unexposed individuals. See text for details. Source: DOE.

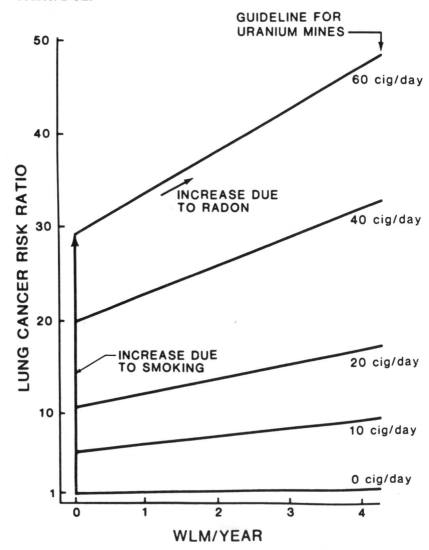

regardless of concern about radon. After all, 150,000 fatal lung cancers per year pretty well attest to the dangers of smoking.

Table 4.2 summarizes some of the findings of the National Academy of Science for lung cancers due to radon. It is important to note that not only uranium miners are affected, but also miners of tin, iron, fluorspar, niobium, and others—all due essentially to poor ventilation and a lack of awareness of the radon problem.

Thus, the uranium miner studies do not unequivocally give much assistance in addressing the question of whether low levels of radiation (in this case radon daughters) have an adverse effect on health. Nevertheless, many workers still argue that the high dose (which may or may not be augmented by smoking) data can be extrapolated to the origin, i.e., zero dose = zero health effects.

TABLE 4.2

Cause-Specific Risks of Mortality among Miners Exposed to Radon Daughters

Study	Observed	Lung Cancer Expected
Colorado Plateau Uranium Miners	185	38.4
Ontario	119	65.8
Elliot Lake Ontario Uranium Miners	81	50.0
Bancroft Ontario Uranium Miners	30	12.4
Eldorado and Ontario	33	7.2
Eldorado-Port radium underground		
surface 2816.0	55	14.7
Eldorado, Beaverlodge		
underground	84	30.0
surface	28	30.8
uranium miners	65	34.2
Swedish iron miners		
Malmberget	50	14.6
Norwegian niiobium workers	12	3.0
Newfoundland fluorspar miners		
(rate ratios)	65	6.5
Cornish tin miners underground		
surface	28	—
Czechoslovakian uranium miners	115	42.0
British iron ore miners	69	27.0

Source: National Academy of Science 1988.

The linear (or linearity) hypothesis determined adverse health effects at low dose. This is shown in figure 4.8. Curve A assumes such a linear relationship. Curve B is a quadratic curve, where it is assumed that the risk decreases more rapidly than the dose. Curve C is a compromise, i.e., a linear-quadratic relationship that advocates more linearity than the quadratic case but less than the linear case. All these curves are speculative for low levels of radiation from radon daughters. Further, the problem of HORMESIS, as well as the problem of threshold, are not considered. Hormesis assumes that for many substances, adverse health effects occur at low dose as well as at high dose and that there is some safe zone at a low, but not the lowest, level. An example would be selenium, where selenium poisoning at high concentrations is well known, yet at very low concentrations selenium deficiency causes muscular problems and stunted growth. It is not clear if the low levels of radiation

FIGURE 4.8
Projection from high-dose to low-dose health risks for radiation exposure. See text for details. Modified from National Academy of Science (1988).

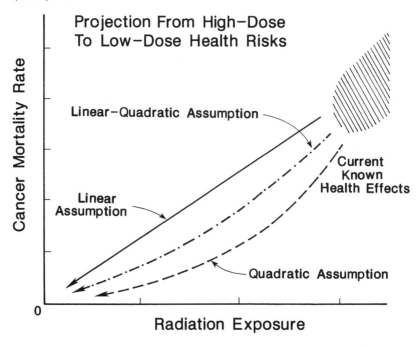

received by all living things are actually beneficial, but there is a large body of data that argue for radiation at least not being unduly harmful (see the study on People's Republic of China in chapter 11). The matter of threshold is also relevant here. Our natural environment is radioactive, and we are subjected to background radiation of different amounts all the time. Statistics do not support the hypothesis that continued exposure to above background levels of radiation are necessarily harmful. Thus, a threshold, but not necessarily hormesis, is likely for radon and its daughters.

So, now back to radon. The assumption is made that exposure to air containing high concentrations of radon and radon daughters has an adverse effect on health, with lung cancer the result in the higher exposure incidences.

Figure 4.9 shows the data for many different miners (mentioned earlier in this section) as well as three very different estimates of cases of excess lung cancer per WLM-year (times 106). One curve assumes the linear hypothesis and places high weight on limited data for high dose-recipient miners, low weight (evidently) to the nonsmokers, and the intercept on the ordinate is 38. The BEIR report (1972) estimates this intercept is 6.5, and Cohen (1979) estimates it at 10—both figures well below 38. Inspection of the data clearly shows that the linear hypothesis curve does not fit the data well at all. Either the BEIR or Cohen value is more realistic.

Thus the data on miners, even including uranium miners, are ambiguous in this matter (NAS 1988). There is, however, another way of understanding the adverse health effects of excess radon. There are five major kinds of lung cancer: epidermoid, small cell undifferentiated, adencarcinomas, large cell, and mixed. For the general population, the small cell undifferentiated lung cancer is only 10 percent of the total, yet for uranium miners this figure may be as high as 70 percent. This indicates that there must be a positive correlation between small cell undifferentiated lung cancer and radon dose. Thus, out of 180,000 lung cancer fatalities in the United States per year, if the 10 percent small cell undifferentiated are due to radon, then radon deaths per year equal 18,000. Interestingly, this 18,000 fatalities per year figure falls in the range of 5000 to 30,000 suggested by EPA, DOE, and NAS. The NAS has found that this radon and small cell undifferentiated cancers relationship may not be as clearly defined as suggested by Cohen, however.

FIGURE 4.9
Excess lung cancer incidence versus radon exposure. See text for details. Modified from National Academy of Science (1988) and Cohen (1979) and others. The dots are mean values for various studies and the bars about each point are the ranges of uncertainties. The dashed curve (Archer et al, 1978) does not fit the data, and the BEIR (Biological Effects of Ionizing Radiation) and the curve marked 'used here' give better estimates.

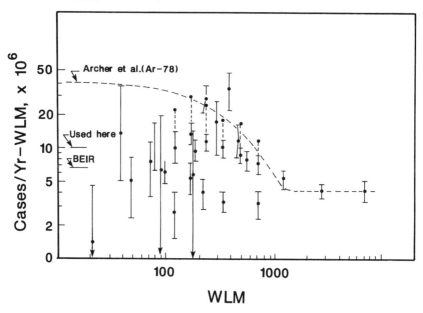

There is another aspect of radon exposure that receives very little attention. It is well known that the average person in the United States receives about 150 to 200 mrem of radiation each year from naturally radioactive materials in the ground, cosmic rays, food and water, and diagnostic x-rays as well as minor radiation from coal-burning power plants and small amounts from above-ground nuclear weapons use and testing and nuclear reactor emissions. These data are summarized in table 4.3. Yet, on a routine basis, radon is not usually included. This is strange, for a person living in a house with 3 pCi/L radon is exposed to about 300 mrem of radiation each year, i.e., greater than the other natural sources mentioned in table 4.3. The National Council on Radia-

TABLE 4.3

Radiation Doses to Individuals in the United States[a]

Source	Millirems
Radioactive elements in ground	26
Food and water (^{40}K, ^{14}C mainly)	28
Cosmic rays (sea level)	28
Cosmic rays (assumed elevation 4000 feet)[b]	(68)
Diagnostic x-rays	70
Fallout from nuclear weapons	4
Nuclear power plant emissions	0.3
Coal power plant emissions	0.4

[a] This table does *not* include radiation from ^{222}Rn and daughters.
[b] Cosmic ray flux increases with elevation; add 1 mrem/100 feet elevation above sea level; i.e., for 4000 feet add $40 \times 1 = 40$, then add to 28, $= 68$.
Source: Toohey, 1987.

tion Protection (NCRP) estimates that the average citizen in the United States receives roughly 100 mrem/year to the lungs from radon; yet this estimate assumes background radon levels of only 1 pCi/L or less. For higher radon levels, as indicated above, the dose to the lungs increases dramatically.

The issue described above is not just an "apples and oranges" type of comparison. Statistically, natural background radiation can lead to a small number of cancers per year. Natural background radon can similarly lead to cases of lung cancer, with the risk being proportional to the radon level.

The EPA and others note that while the risk of lung cancer from radon exposure is low for low radon levels, it is not "zero." For houses with 4 pCi/L and an assumed residence time of 50 years, the risk of dying from lung cancer is two percent, assuming that the linear hypothesis holds at low levels of radiation. For higher levels, then, the risk is greater. The EPA has compared risk of indoor radon in terms of risk from other sources (see chapter 11). Thus, a nonsmoker's risk of dying from lung cancer occurs between one and two pCi/L. At about 30 pCi/L, the risk is comparable to smoking two packs of cigarettes per day.

Toohey (1987) presents a progress report on research into the radon problem being carried out at Argonne National Laboratory. According

TABLE 4.4
Newly Proposed Background
Radiation Sources

Source	Percent of Total
Radon	40
Medical	17
Internal	15
Gamma	15
Cosmic[a]	12
Other[b]	1

[a] Assumed sea level value. Since cosmic ray flux will increase with elevation, this figure will increase in areas of the United States above sea level.
[b] Nuclear weapons fallout, emissions from nuclear and coal power plants.
Source: Toohey 1987.

to him, radon is the major source of background radiation, outweighing the external, internal, cosmic, and even medical components in all cases by a factor greater than two. This is shown in table 4.4.

In addition, Cothern (1987) of the EPA has presented another version of background radiation that includes indoor radon and drinking water radon. His chart is shown as table 4.5 where it is noted that cosmic rays are "averaged" at 45 mrem/year, which would indicate mean elevation of 1900 feet for the conterminous United States. There are other small variations, but it is significant that by including indoor radon, which accounts for 90 mrem per year, the overall approximate total is 300 mrem/year, well above the 150–200 mrem/year usually advocated. The radon figure is apparently based on an assumed indoor value of 0.5 pCi/L, which, in view of the many data now becoming available, may be too low by a factor of 2 or 3. If this is true, then the average indoor- plus drinking-water radon value (table 4.6) may well be in the range of 150–200 mrem/year alone! In other words, an average individual in the United States receives a possible 400 or so mrem/year. This is an extremely important statement, part observation and part speculation, but, if substantiated, it means that our background radiation may be much

TABLE 4.5
Natural Background Radiation

Source	Dose (mrem/year)
Natural	
Cosmic rays	45[a]
External sources	40
Internal (food, drink)	20
Radon	90
Anthropogenic	
X-rays	80
Radiopharmaceuticals	16
Fallout	3
Nuclear Power Plants	0.1[b]
Color TV	1
Mining and milling uranium and phosphate rock	5
Occupational exposure	0.8
Consumer products	0.3
Approximate total	300

[a]The cosmic ray value must assume 1900 feet mean elevation; i.e., base of 26 mrem/y at sea level plus 1 mrem/y for each 100 feet elevation.
[b]Coal power plants add about 1.5 to 2 times this amount.
Source: Cothern 1987.

TABLE 4.6
Rn Concentration in Water Versus Lifetime Cancer Risk

Radon Concentration (pCi/L)	Lifetime Risk (%)
5,000	0.2 to 0.4[a]
10,000	0.4 to 0.8
15,000	0.6 to 1.2
20,000	0.8 to 1.6
25,000	1.0 to 2.0[b]

[a]Equivalent to estimated lifetime risk from estimated lifetime average indoor radon concentration of 1 pCi/L air in houses.
[b]Equivalent to estimated lifetime risk of occupational standard of 4 WLM/year for 30 years.
Source: Cross et al. 1987.

higher than we have previously assumed. Hence, the use of the linear hypothesis to low levels of radiation is even less justified since the numbers of adverse health effects (i.e., cancers) have not been realized. This will be discussed further in chapter 11.

NONRADON LUNG DISEASES OF URANIUM AND OTHER METAL MINERS

The National Academy of Science (1988), in its comprehensive study of indoor radon risk, also addresses other causes of lung disease in uranium miners and in other metal miners. A significant correlation of emphysema, for example, is noted in uranium miners, and pulmonary fibrosis and silicosis are also identified. Other metal miners show elevated levels of these diseases as well. Data for these diseases are not as well characterized as the lung cancer deaths from radon and radon daughters, however. The smoking history of the individuals exhibits a multiplicative relationship with emphysema, fibrosis, and silicosis.

Miners are hardy and usually do not wear respirator masks while working, especially when low levels of radon have been monitored. This increases their likelihood of inhaling siliceous material. In addition to the emphysema, fibrosis, and silicosis, miners are often subject to high incidence of chronic bronchitis and pneumonia.

Miners are often exposed to toxic fumes from power and transportation equipment, chemicals, heavy metals, impure water, and many other factors, which make mining a truly risky occupation.

URANIUM MILL TAILINGS

The National Academy of Science has also investigated uranium mill tailings in great detail. I served on this panel from 1984 to 1987 until their report was published (NAS, 1987).

Much, if not most, of the concern about uranium mill tailings focuses on radon emanations from these waste piles. Twenty-five inactive uranium mill tailings sites are being assessed for actions to render them as safe as possible; these sites, located mainly in the western United States, except for a site at Canonsburg, Pennsylvania, represent some 25 million metric tons of wastes. Sites that are active or on standby are much larger and total another 65 million metric tons of wastes. As pointed out earlier in this book, the milling of uranium is not 100 percent effective, and the tailings contain above-background uranium as well as ^{230}Th and ^{226}Ra. The long half-lives of ^{230}Th (80,000 years) and ^{226}Ra (1,625 years) means that they will be available to generate radon for close to a million years, and the uranium for much longer periods.

However, the tailings problem must be put into perspective. The NAS concluded that the risk from radon to the average citizen of the United States is so low as to be trivial. If we assume individuals live on the edges of tailings, then the risk may be locally higher.

The NAS also reports that radon having migrated away from most tailings piles is the same as background levels less than 0.5 kilometers away, and even in areas of high topographic relief, such as canyons, etc., the radon reaches background in a kilometer or so.

Interestingly, the real risk from uranium mill tailings lies in potential pollution of groundwaters from nonradioactive sulfate, chloride, nitrate, some organics, and various metals. The risk from radioactive materials is, indeed, very slight in comparison.

Risk from radon in terms of annual fatalities has also been assessed. The radon released from soil tillage is assigned a fatality figure of 1150 cancers; from building interiors, 1600 cancers; from fertilization, 0.5 cancers; and from mill tailings, somewhere between 0.02 and 2 cancers, assuming an individual lives directly over the tailings. These figures should be compared with the 20,000 fatalities estimated by the EPA and the 15,000 estimated by the NAS from indoor radon! To illustrate this, I have mentioned to my students that they would receive less radon sleeping at the fence line surrounding uranium mill tailings than if they slept in their own bedrooms in Albuquerque.

I do not intend to belittle the seriousness of uranium mill tailings, but rather to point out that in terms of relative risk to the average citizen in the United States, the tailings radon risk is low indeed. Further, uranium mill tailings are regulated, they are monitored, the problems, present and future, have been identified, and a vigorous program is underway to mitigate these tailings as necessary.

RADIUM HEALTH EFFECTS

Radium (^{226}Ra) is the immediate parent to ^{222}Rn, and it is also highly radioactive and a known carcinogen. The EPA has taken the position that radon inhaled is a more serious threat to public health than radon taken in by ingestion of food or water. Recently, Dr. Douglas Mose of George Mason University, Virginia, has been compiling health statistics along with well-water radon contents for areas in Maryland and Virginia. While his data are still in the preliminary stages at the time of this writing, they nevertheless show a striking correlation of high water radon content with incidences of digestive track cancer. His suggestion is that the radon in the waters is proportional to the parent radium and that the cancers may in part be caused by the radium. This implies that by monitoring waters for radon content, which is fairly easy and straightforward to carry out, one may obtain indirect data for the more potentially dangerous radium, which is very difficult to measure in waters.

CONCLUSIONS

Fatal lung cancers may result from exposure to radon and its decay products. Estimates of such fatal lung cancers per year in the United States range from 5,000 to 35,000, although the National Academy of Science (1988) now estimates 15,000 to 20,000. Thus, indoor radon is the second leading cause, behind smoking, of fatal lung cancers per year in the United States. In chapter 11 some comparative risk estimates are made, and it is noted there that of fatalities per year from natural causes indoor radon is the leading cause .

In this chapter, I have briefly examined how radon and its decay products enter lungs and cause damage. In the subsequent chapters, I will be more concerned with how and when radon exposure occurs, and what can be done to prevent its build-up in dwellings.

5

Radon Entry: Sources and Mechanisms

As we will see in this chapter, radon entry into structures is both predictable and complex—predictable because of radon's wide distribution; complex because we do not fully understand the mechanisms responsible for radon movement in various types of earth and other materials. The radon budget in rocks, soils, water, and the atmosphere has been discussed in chapter 2 and will be briefly commented on here. The rocks and soil over which structures are built are the primary sources of most radon detected in homes. Secondary sources include building materials, waters (especially well waters) in some instances, and others. In this chapter, I will discuss the role of each of these factors.

SOILS

Weathering of rocks or other earth materials produces a residual end product that is called soil. This soil may be well developed and in a more or less steady state, in which case it is said to be mature. If not, it is referred to as immature. Soils are a function of a very large number of factors, including nature of parent rock(s), climate, vegetation, topography, organic matter (living and dead), and time. Material may be brought in to the soil-formation area from extrinsic sources by wind or water

FIGURE 5.1

Zones in a typical soil. The A horizon is the zone of leaching, the B
horizon is the zone of accumulation, and the C horizon is the zone where
active attack on bedrock or other parent material occurs. Some of these
zones are more susceptible to radon emanation than others; see text for
details.

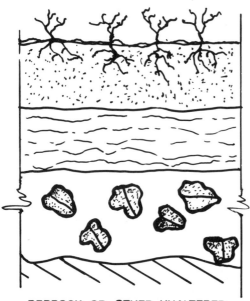

A- HORIZON (TOPSOIL)

ZONE OF LEACHING

B- HORIZON (SUBSOIL)

ZONE OF ACCUMULATION

C- HORIZON

ZONE OF ACTIVE ATTACHME
ON PARENT MATERIAL

BEDROCK OR OTHER UNALTERED
PARENT MATERIAL

action as well. A generic soil consists of three horizons; *A horizon, B
horizon, and C horizon* (see fig. 5.1). The A horizon is the zone of
leaching; here downward-moving water leaches away soluble salts and
colloidal material. In the B horizon, much of the material leached from
the A horizon is trapped and thus is known as the zone of accumulation.
Here clay minerals may form, along with silica, carbonates, and iron
OXYHYDROXIDES. The underlying C horizon is the zone of active
attack on bedrock. The A horizon is usually rich in organic matter to
some degree, with humus causing dark layers in the upper part and a
bleached layer beneath due to leaching. The B horizon is more indurated,
and in the C horizon fragments of bedrock are enclosed by soil.

Soil, then, is a very complex material, and it varies in both physical and chemical properties to a very high degree. Soil may be considered a mixture of rock and mineral fragments and/or organic matter and liquid and/or gas. The gas present in soil is very much like air, except that it can contain high amounts of gases such as radon. The liquid in soils is assumed to be water.

The purpose here, however, is to address those factors that are important to radon migration. Despite the complexity of soils, a few general properties are important to all soils for radon movement. These include porosity, permeability, moisture content, grain size, and mineralogy.

The porosity of a soil refers to its volume of pore space in terms of a percentage. For soils, the porosity varies over wide ranges, as shown in table 5.1. There is also a relationship between grain size and porosity (see table 5.1), because packing sand grains together is a geometric process and a predictable pore space results for spheres of different averages or mixed sizes. For clays, however, there are complex chemical and physical interactions involved that result in a higher overall porosity. Silt, intermediate between sand and clay in grain size, is thus intermediate in porosity as well.

Moisture content of soils varies as a function of time, composition, and grain size. Water may be held in soils in large pores, as films on and around grains, in narrow channels, loosely fixed in clay minerals and in some oxyhydroxides; and as water of crystallization for some minerals. Pore water is strongly affected by gravity, but the other sources may persist for long periods of time. A soil is saturated when the total amount of moisture equals the porosity.

But moisture content and porosity alone may not necessarily cause migration of liquid or entrapped gases. The ability of fluids to move through a medium is called its permeability. Typical permeability values for clay, silt, and sand media are given in table 5.1. Soil permeability is

TABLE 5.1
Soil Properties as a Function of Grain Size

Material	Grain Size (microns)	Porosity	Permeability (m^2)
Sand	60–2000	0.4	4×10^{-11} (grain size = 200 μm)
Silt	2–60	0.5	1×10^{-12} (grain size = 20 μm)
Clay	2	0.6	1×10^{-14} (grain size = 1 μm)

probably the single most important soil property for radon migration. The higher the permeability, the greater the potential for radon movement. For layered rocks, such as sandstones and shales, and for layered soils, there is a tendency for a greater permeability parallel to the direction of the layering as opposed to right angles to the layering. Thus, for flat lying strata, horizontal permeability is greater than vertical permeability. This, for radon purposes, means that a source of radon well removed from a structure may migrate toward that structure by horizontal permeability. Obviously, if there are fractures in the layered strata or vertical channels in flat lying soils, then there is a greater potential for vertical migration as well. For example, if the foundation of a dwelling is blasted into rock, and this blasting and other construction cause some fracturing of the underlying media, the result may be favorable radon migration up the new fractures.

Permeabilities may vary over several orders of magnitude for various earth materials (see table 5.1) and within any particular soil, as well as being a function of mineral matter, organics, channeling, etc. Chemical reactions are common in soil, and many such reactions affect the permeability of water flowing through the soil—but this is not necessarily the case for the inert gas radon. Hence, even in cases when permeability starts to decrease due to chemical reactions, the radon permeability may not decrease.

Some mineral reactions may have important long-range effects on radon. In arid climates, calcium carbonate or calcium sulfate is commonly fixed as "caliche" in soils (respectively called calcrete or gypcrete depending on whether calcite or gypsum is formed). These caliche layers are commonly highly impermeable and act as traps for migrating uranium. The western United States contains, for example, numerous small economic uranium occurrences of this type as well as many more occurrences where, while not ore grade, uranium is nevertheless enriched. Radium, too, remains immobile in these settings. The concern is that when areas are leveled for building, whether on a local or a large scale, this impermeable barrier is commonly disturbed, thus causing potential for radon migration from the caliche zones into overlying dwellings.

Two important processes for permeability are diffusion and convective flow. Diffusion is the process by which a particular species moves in response to a chemical gradient. Convective flow means that a circulating system of fluid causes matter to migrate. Both processes are important by themselves and in combination. Near buildings, radon movement is commonly diffusion-dependent; removed from buildings, convective flow may be more important. Diffusion is the dominant process by which the earth's radon is lost to the atmosphere on a global scale. More detail on porosity, moisture content, and permeability aspects of soils can be found in the readings suggested for this chapter.

Diffusion is a function of porosity, moisture content, grain size, and permeability. The upper limit for radon diffusion is set for open air, which is 10^{-5} m²/s (meters squared per second), while in many sands with variable clay content, values on the order of 1 to 3 × 10^{-6} m²/s are found. In some clays and muds, values from 10^{-7} to 10^{-10} m²/s are noted. The effect of different amounts of waters on radon emanation is discussed later in this chapter.

Radium in Soils. The immediate radioactive parent of ^{222}Rn is ^{226}Ra. In soils, radium is transported as a simple ion, Ra^{2+} in solution. It is insoluble in the presence of moderate amounts of dissolved sulfate (SO_4^{2-}) and carbonate (CO_3^{2-}) and will form (usually) radium sulfate or (less commonly) radium carbonate. Often the Ra is coprecipitated with barium in the mineral barite, (Ba, Ra) SO_4 or in calcite, (Ca, Ra) CO_3. The soil radium is largely controlled by the radium content of the source rocks. Some typical values of radium in different rocks are given in table 5.2 for ^{226}Ra. Radium from thorium, ^{228}Ra, the parent to so-called thoron, ^{220}Rn, is not considered here (see discussion in chapter 2). More detail on radium in soils and rocks is given in chapter 2 of this book and in Nazaroff et al. (1987).

Release of Soil Radon. Radon in soil correlates with its immediate radioactive parent, ^{226}Ra, with ^{230}Th, and to a lesser degree with its ultimate parent, ^{238}U. From the decay chain of ^{238}U, (fig. 3.1), it is apparent that there are several places for loss of intermediate daughter products to take place. The ^{238}U in many rocks and minerals occurs in sites compatible with U^{4+}. However, in the process of decay of ^{238}U to ^{234}Po and then to ^{234}U, the energetics of decay cause oxidation of insoluble U^{4+} to U^{6+}, which is more easily lost from rocks and minerals. If the U^{6+} is carried in solution, some precipitates when ^{230}Th is formed (i.e., Th^{4+}, a highly insoluble form, is the only naturally occurring valence). This, in turn, decays to ^{226}Ra and then to ^{222}Rn. If such losses

TABLE 5.2

Radium Concentration in Rocks

Rocks	Number Samples	^{226}Ra (pCi/Kg)
Rhyolites	131	1,900
Granites	569	2,100
Andesites	71	700
Diorites	271	1,100
Basalts	77	300
Gabbros	119	300
Dunites	31	800
Phonolites	138	9,900
Syenits	75	18,700
Nephelinites	27	800
Foidites	8	800
Evaporites	243	1,200
Limetones	141	700
Sandstones	412	1,600
Clays	40	1,400
Shales	174	2,000
Coglomerates	198	1,400
Gneiss	138	1,400
Schist	207	1,000

Note: The range of data for each rock type may be very large. The mean value thus has an associated +20% error of the mean.
Source: Wollenberg 1984.

occur, then the remaining uranium in a soil may not be in equilibrium with radium present, but the radium will be much more closely in equilibrium with radon. Further, if this scenario is correct, much of the thorium-radium-radon material is located on grain surfaces, in fractures, and other defects in minerals such that percolating waters can more easily promote radon loss than can be promoted in the original U^{4+} sites, where the radon is more tightly locked in. That this must be the case is further documented by the fact that minerals dated by the U-Pb method (see chapter 2) usually give fairly reliable dates; thus, the loss of intermediate radioactive uranium daughters must be minimal for such minerals.

In soils, several factors are important for radon release. The process

FIGURE 5.2

The role of alpha recoil on radon emanation in soils. If water is present the recoiled particle is trapped between grains. If air is present, some of the recoiled particles will enter the adjacent grains as shown (6.2b). The entry pathway may be later more readily altered. See text for detail. Modified from R. L. Fleischer.

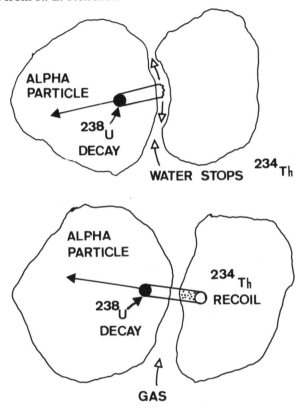

of alpha recoil is shown in figure 5.2. In this process, an alpha particle is given off within the grain, and, since this is a high energy reaction, the new isotope ^{234}Th is formed. This isotope is recoiled in the opposite direction. If, as drawn in figure 5.2a, the recoil ^{234}Th enters intergranular water films, it is stopped and will be "flushed out" of the original volume where formed. In figure. 5.2b, however, gas is present between the two grains. In this case, the recoil particle, ^{234}Th, will not be stopped by the intergranular medium, but will rather impact on and into an adjacent mineral grain. The entry passage of this recoil particle in the new grain is weakened, as indicated in figure 5.2b, and thus, when radon is eventually formed here, it has a higher likelihood of migrating to the grain surface and into gas or air between the grains. Similarly, the alpha recoil of the decay from ^{226}Ra to ^{222}Rn is important. The range of ^{222}Rn transport due to this recoil is about 60 microns in air, about 0.006 microns in water, and much less within mineral grains. Stated simply, in the absence of intergranular films of water, the ^{222}Rn lost by recoil may actually be taken up by another mineral grain. On the other hand, if water is present, this radon will be dissolved. The diffusion of ^{222}Rn is orders of magnitude different in separate minerals. In air, atoms diffuse at about 10^{-5} m^2/second, and in solids from 10^{-12} to as little as 10^{-24} m^2/second. Diffusion rates in water are intermediate between those in air and solids. Even when the alpha-recoil entrapped radon is not lost initially, the damage to the mineral grain is more susceptible to alteration; hence, ^{222}Rn entrapped there can more readily be lost relative to that still locked into uranium sites (see fig. 5.1).

For radon released from minerals and incorporated into intergranular waters, as the water table is met, then radon is lost to pore space in the vadose zone from which it is emanated to the atmosphere or into structures above the sites of release.

Quantitatively, the emanating power, P, for a soil is a ratio of radon atoms that escape into surrounding pore space relative to the total number of atoms of radon formed by decay from ^{226}Ra. The production rate, R, per unit volume of pore space in the soil is:

$$R = \frac{P\, n_t\, p}{\theta}$$

where P = emanating power, n_t = total atoms of Rn formed from Ra, p = density, and θ = porosity (see Wilkening 1985).

Radon lost from minerals will readily be lost from pore spaces, especially in the vadose zone, where mineral grains are unsaturated but often contain thin films of water. As saturated conditions are encountered, radon is more effectively kept in the water. Some experiments show, for example, that during its mean life, radon can diffuse from one meter of dry soil but only one centimeter from water saturated soil. For lightly coated grains, the thin water film prevents the dissolution of alpha-damaged mineral sites. In the absence of water, alpha-damage is promoted, and thus radon loss is also promoted. The result of these processes is an overall radon flux to the atmosphere of 0.6 pCi m²/s. Nero (1987) estimates the average radon content of the atmosphere at 0.1 to 0.4 pCi/L and also notes that this may account for 10 to 20 percent of indoor radon in many structures.

The average soil radon content is not well defined. However, Terradex, a subsidiary of Tech/Ops, has made the most radon measurements. Using soil alpha track etch detectors, the firm (R. Oswald, written communication, W. Alter, oral communication) suggests that at a depth of 38 centimeters (15 inches), typical soil in the United States averages about 100 pCi/L. This is based on over 500,000 determinations. Very simply, there exists in soils, worldwide, a sufficient reservoir of easily available and mobile radon to enter virtually any structure built anywhere.

COMMON DWELLING STRUCTURE TYPES: UNITED STATES

Dwelling structure types for single families are, of course, extremely varied. For the sake of simplicity, I will briefly discuss several generic types here. Many dwellings are constructed on grade with soil, while others are constructed with subgrade lower limits and some are constructed above grade. Basements and subgrade, first-floor, split-level homes are good examples of subgrade structures, while wood houses are commonly constructed above grade over a foundation. The space between the grade surface and the bottom of the floor is the crawl space. Homes built on grade commonly have a cement slab underlying the floor.

Much research has been focused on homes with basements since these are, in many instances, the most vulnerable to radon entry. Figure 5.3 shows some ways in which radon may enter a hypothetical home. It is clear that many (if not most) of the entry paths are in the basement walls, floor, etc. This is not surprising because soil radon gas is commonly quite high (see discussion earlier in this chapter), with average soil gas (in the United States) at about 100 pCi/L. In some areas of the United States, such as these overlying granitic debris, average soil radon gas may be much higher. I have reported (1977) an average of about 180 pCi/L for soil radon gas in the Albuquerque, New Mexico, area, for

FIGURE 5.3
Ways in which radon may enter a house. See text for details. From EPA.

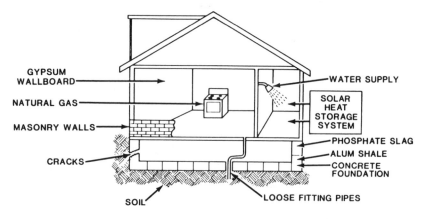

GYPSUM WALLBOARD

NATURAL GAS

MASONRY WALLS

CRACKS

SOIL

WATER SUPPLY

SOLAR HEAT STORAGE SYSTEM

PHOSPHATE SLAG

ALUM SHALE

CONCRETE FOUNDATION

LOOSE FITTING PIPES

example (to be discussed in detail in chapter 7), although my recent (1988) data suggest the soil radon average is higher. The soil radon gas exerts a positive radon pressure against the subgrade walls and leads to penetration of radon gas in areas of defects.

Houses made of wood are commonly constructed above grade, with a foundation and crawl space separating floor from soil surface. Like subgrade basements or crawl spaces, the crawl space for the wood house may also be a potential source of high radon. This is especially true since, although the foundation walls may be tight, the soil is left uncovered. Hence radon emanating directly from the soil can accumulate in the crawl space.

Houses built on grade are usually built over a cement slab, which, in turn, is usually built over gravel fill that may or may not be sealed. Sealants are commonly used to prevent moisture from entering the cement, but they are effective in preventing radon from entering as well. In actuality, a good, tight cement slab is an effective barrier to much radon entry. However, cracks can form in slabs, and pipe or other openings may not be tightly sealed enough to prevent radon entry.

Two factors dominate potential radon entry into houses: how effectively the house is coupled to the soil and sizes of openings and penetrations (for a thorough discussion of this topic, see readings for this chapter). The crawl spaces for wood houses may be inconsequential for radon accumulation if they are well vented; if, on the other hand, they

TABLE 5.3

Some Common Openings for Radon Entry into Homes

Cracks between floor and walls of basements
Cracks between blocks or bricks in walls in basements
Openings around pipes, outlets, wiring, etc.
Openings around anchor bolts for pumps, equipment, etc.
Seals around drains
Construction flaws or corrections (plumbing mistakes, etc.)
Sunken rooms (for on-grade homes)
Subsidence cracks
Miscellaneous shrinkage cracks
Hollow cement (cinder block) walls
Openings between floor sections
Intentional openings (i.e., drains)

are sealed, as is commonly done during winter months, then radon can easily build up in the spaces. For ongrade and subgrade houses, the potential for radon gas accumulation may be greater than that for crawl spaces because the pressure of radon gas in the soil is so much higher than that inside the dwellings.

Scott (1987) has pointed out the importance of careful construction for all types of dwellings. When basements are built, the walls are first built, then the basement floor is poured while the walls are still wet. This often causes cracks to develop between the basement floor and walls, which is an easy opening for radon gas to penetrate. Other openings for radon penetration include those around pipes, construction flaws, wiring conduits, pump outlets, special architectural features, and others. These are given in table 5.3.

RADON FROM WATER

Radon can also enter a dwelling through water. Nazaroff et al. (1987b) have summarized water aspects of radon in domestic waters from the findings of many investigators, and their work should be considered for an in-depth treatment of the topic. In the 1950s, high radon levels were first observed in well waters in Maine. While the first concern was about radiation due to ingestion of these waters, it was later argued that the radon released in gaseous form from the waters might be of more concern. More recently, the National Academy of Science (1988), citing the works of others, has suggested that the risk to lungs from inhalation of radon from potable waters is three to twelve times that of stomach cancer from ingested radon.

Radon is soluble with some difficulty in water and is easily lost when the water is warmed. Nazaroff et al. (1987) summarize a wide variety of studies on fractions of radon lost from different water sources: dishwasher, 0.95; shower, 0.66; bath, 0.42; toilet, 0.3; laundry, 0.92; drinking and cleaning, 0.34. These results show clearly that more radon is lost from hot water sources such as dishwashers and laundry use, with somewhat high values from showers and baths), and lowest release from toilets and drinking and cleaning. Problems in obtaining data thus arise since many studies involve testing of one but not several of these possible uses. Hence, radon-in-water data are subject to large errors in many instances.

The United States public derives its water from surface sources (49.5 percent), public groundwater supplies (33.2 percent), and private wells (18.3 percent). Public groundwater radon values for 41 states are given in table 5.4. Values for radon concentrations range from 33 pCi/L to 1773 pCi/L for the means (see table 5.3). Radon contents as high as 27,000 pCi/L have been measured. A key question is, what effect on indoor radon in the air does the radon from water have? In table 5.5 the data for average waters in the United States are summarized . Surface waters yield an average of only 0.0005 pCi/L, groundwaters 0.009 pCi/L, and well waters 0.06 pCi/L. At first glance, then, radon from all waters appears not be of too much importance to the indoor radon problem. Yet, in areas where private wells are common and where certain uraniferous and/or radium-rich rocks are in communication with

TABLE 5.4
^{222}Rn Abundances in Public Groundwater Supplies in the United States[a]

State	Population Served (thousands)	No. Samples	^{222}Rn (pCi/L)
Alabama	1,200	104	89
Arizona	1,490	64	295
Arkansas	880	43	46
Colorado	320	37	252

State	Population Served (thousands)	No. Samples	^{222}Rn (pCi/L)
Delaware	254	36	96
Florida	6,800	165	82
Georgia	1,320	61	105
Idaho	592	85	224
Illinois	4,050	158	134
Indiana	1,920	117	68
Iowa	1,600	58	118
Kansas	9,033	7	72
Kentucky	375	50	94
Louisiana	1,850	22	93
Maine	101	68	1,011
Massachusetts	1,550	100	724
Minnesota	1,910	124	182
Mississippi	1,800	53	57
Montana	184	33	378
Nebraska	961	21	178
Nevada	329	26	359
New Hampshire	392	31	868
New Jersey	3,420	19	365
New Mexico	798	89	203
New York	3,510	1,500	110
North Carolina	474	181	132
North Dakota	2,258	67	121
Ohio	2,950	84	120
Oklahoma	662	56	130
Oregon	344	65	228
Pennsylvania	2,180	89	384
Rhode Island	142	92	1,773
South Carolina	541	185	136
South Dakota	321	79	266
Tennessee	1,450	50	34
Texas	5,030	278	131
Utah	662	98	786
Vermont	113	11	624
Virginia	707	101	226
Wisconsin	1,620	143	201
Wyoming	122	18	343
Total	56,085 [a]	3,318	140 [b]

[a] Includes 76 percent population served by public groundwater supplies.
[b] Population-weighted statistics.
Source: Modified from Nazaroff et al. 1987b; data are gross means.

TABLE 5.5

Indoor United States airborne ^{222}Rn: Common Water Types

Type	Fraction of United States	Water-derived Indoor Rn(pCi/L)
Surface	0.496	0.0005
Public groundwater	0.322	0.009
Private wells	0.06	0.06

Source: Solly et al. 1983; Nazaroff et al. 1987.

the wells, the release of radon from well waters may be quite high. As a rule of thumb, a water radon content of 10,000 pCi/L may release an equivalent 1 p Ci/L to indoor air, although this assumes a high degree of released radon gas (see table 5.3). Surface waters, on the other hand, contribute an insignificant amount of radon to air in dwellings, and even public groundwaters amount to only 0.8 to 2 percent of total indoor radon. Remedial action for radon released from high water sources is covered in chapter 8

FIGURE 5.4

Map of the United States showing locations of groundwaters with different radon levels. See text for details.

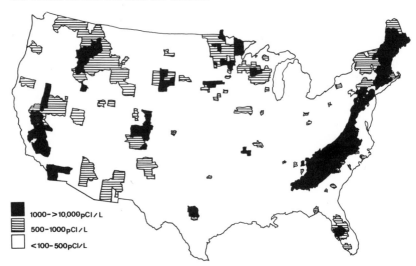

1000->10,000pCI / L

500-1000pCI / L

<100- 500pCI/L

A map of the forty-eight contiguous states that gives the radon concentrations in groundwaters is shown in figure 5.4. For convenience, the waters are divided into waters with less than 500 pCi/L, 500 to 1,000 pCi/L, 1,000 and 10,000 pCi/L, as well as those over 10,000 pCi/L of radon concentrations. In general, the radon in groundwater pattern mimics the distribution of radium in the same waters according to Michel and Jordana (1987). Geologic control is evident for much of the distribution. The core rocks of the Appalachians, from Maine to Alabama, are pervasively high in radon. So, too, are areas with sediment derived from granitic terrains, as in parts of California, Colorado, Idaho, and New Mexico. In Florida, however, there is a pronounced radon high surrounding Tampa Bay which is due to the high uranium- and radium-bearing phosphate rock found and processed in the area.

RADON FROM NATURAL GAS

Many natural gases contain high quantities of dissolved radon at the well head. Much of this, however, is presumably lost during transport, and the amount of radon released to indoor is assumed to be insignificant.

AEROSOLS AND ATTACHMENT OF RADON DAUGHTERS

Radon daughters may attach themselves to solids, such as walls, ceilings, floors, furniture, etc. Radon daughters may also be fixed or attached to aerosol particles in the air. The importance of these aerosols is that, while they effectively remove some of the radon daughters, they may act as available sources for excess radon daughters to the lungs if allowed to accumulate (i.e., in houses where dusting is only done infrequently). In testing for indoor radon, most methods use a filter to trap out the radon daughters (both attached and unattached); hence, the amount of radon daughters proportional to radon is a minimal value (see chapter 6 for details). Similarly, detection methods for WLM on radon daughters (chapter 6) may show falsely high results.

While radon is an inert gas, its daughters are commonly charged and chemically reactive. Many of the radon daughters become attached to aerosols, usually soon after their formation, but some remain unat-

FIGURE 5.5

The role of plateout on indoor radon. See text for details. From DOE.

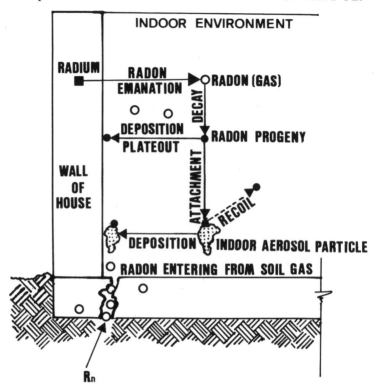

tached. The interactions of radon and radon daughters with indoor aerosol particles is shown in figure 5.5 as a complex series of emanations and removals of radon and radon daughters—again attesting to the problem of across-the-board characterizing of any volume of indoor air with respect to radon and its daughters.

RADON FROM BUILDING MATERIALS

Building materials may not contribute almost negligible to major quantities of indoor radon. In an attempt to use uranium mill tailings for practical purposes, a large number of homes in Grand Junction, Colorado (NAS, 1988), were built using bricks made of the tailings. Not surprisingly, the radioactivity levels in these homes and the indoor radon levels were found to be very high. In the milling of uranium ores, usually well over 95 percent of the uranium is removed. Hypothetically, if an ore contained 0.4 percent uranium oxide before milling, then removal of 95 percent would still leave 200 parts per million (ppm) uranium oxide after extraction, which is well above the average for most rocks (and therefore most building materials). In the Grand Junction case, the homes had to be destroyed and new dwellings constructed.

Thus, building materials can, if not carefully considered, lead to potentially high amounts of indoor radon. Fortunately, most building materials contain relatively small amounts of precursor ^{226}Ra; hence, the release of the daughter ^{222}Rn is also small. Nevertheless, routine monitoring of building materials may be advisable, especially in cases where high indoor radon contents do not apparently correlate with high soil radon, degree of insulation, and other factors.

In Sweden in the 1950s, it was noted that for homes built using the highly uraniferous Alum Shale as an essential ingredient, the radon levels were high. Later, it was determined that probably 10 percent of homes in Sweden have been built using Alum Shale as a cement/concrete component, although its use has been banned for many years. Later studies show that even in these cases, the majority of radon comes from the soil. In the United States, plasterboard made from by-product gypsum from the phosphate industry is commonly uraniferous. This results from the fact that U^{4+} and Ca^{2+} are almost identical in ionic radius; hence, uranium easily substitutes into the main phosphate ore mineral, apatite (calcium phosphate) (see chapter 2). This problem, too, has been known for some time, and most phosphate ores are thus treated for uranium extraction, although, like the uranium ore mentioned above, the waste material still contains above-background uranium. In addition, radium as Ra^{2+} is also incorporated into the wastes. Still further, the raw ingredients in material such as cement have variable uranium and radium contents. I have noted (1977) values of uranium from 2 to 30 ppm for the Madera Limestone, the main ingredient for cement manufacturing in the Albuquerque area. Here there is another possible source of uranium and radium in that by-product ash from a coal Power Plant in Grants, New Mexico is used as the main mixing ingredient with the limestone, and such ashes typically are enriched in uranium over the source coals by very large factors.

Wood structures contain the lowest potential sources of indoor radon.

This is due to the fact that most trees do not readily incorporate either uranium or radium into their cellular structure. I am unaware of any documented case of wood as a major contributor to indoor radon.

This is not true, however, for other building materials. As mentioned above, cement and concrete may contain slightly to moderately (rarely highly) elevated uranium, depending on the local rock geochemistry (see chapter 2). Shale is used as the source of silica, iron, aluminum, and other non-calcium, magnesium carbonate components of cement. Since shales chemistry is quite variable, it is possible that in some areas high uranium (and radium) shales may be used for cement manufacturing. In eastern United States, the Chattanooga Shale is actually classified as a very low-grade uranium resource because of its elevated uranium content. Shales in many other parts of the United States and worldwide may be locally rich in uranium or radium (see discussion in Brookins, Weddington, and Merklin 1981).

Other common building materials, such as bricks, adobe, sandstone, cinder block, slump rock, granite, limestone, and others, may be factors if not monitored. Many granites are moderately high in uranium, for example. In several places in New England, fairly elevated uranium contents are noted (as for Conway granite in New Hampshire). Elevated uranium, radium, and soil radon are common in immature soils developed from granitic terrains, (see Brookins 1988). The radioactivity of such granites is well known. The facings and many structural elements in Grand Central Station, New York City and in the Capitol and Congressional office buildings in Washington, DC, yield moderately high dose rates of radioactivity (see chapter 11).

Adobe is a common building material in the western United States and elsewhere in the world. Adobe is made by mixing sand and clay with straw and water and allowing the mixture to harden. In the Albuquerque area, several of the clay pits used for adobe manufacturing contain slightly elevated uranium levels (up to 40 ppm U; Brookins 1988). It is not known as of the time of this writing if this is a factor of importance. Bricks, too, use clay of variable composition, and the same unknowns as adobe are present here as well.

Cinder block is a mixture of cement with volcanic rock, commonly pumice or scoria. Here the potential sources of uranium are the limestone and shale (or fly ash) in the cement and the volcanic material. Many volcanic rocks contain appreciable uranium, while others are low in uranium content. Again, only by actual analyses can the uranium (or radium) levels be determined.

One of the advantages of the use of materials such as cement, however, is that the raw materials are fired at about 1500°C, a temperature at which water and carbon dioxide are driven off, and the resulting material is partially a molten residuum called clinker. The uranium and radium are often in the glassy fraction of the material, and, as such, are not easily susceptible to loss of any radon formed. This is discussed below.

The radium content of various rocks and other materials is given in table 5.6. It is noted that the values, reported as ranges in pCi/kg material, vary from a low of 135 pCi/kg for naturally occurring gypsum to over 67,000 pCi/kg for concrete made with Alum Shale (Sweden). Fly ash values range from several hundred pCi/kg to well over 50,000 pCi/ kg (Brookins 1988).

Loss of radon from building materials is controlled by two processes, diffusion and flow. The diffusion of radon is controlled by its concentration gradient and takes place in response to this gradient (see earlier discussion in this chapter). In contrast, the radon that moves by flow occurs in the pore space and interstitial parts of the material. Flow is dependent on several factors, including pressure, temperature, porosity, permeability, and moisture content. Just as radon in soil is controlled by

TABLE 5.6
Radium Content of Some Common Rocks and Building
Material Ingredients

Material	Radium Concentration (pCi/kg)
Cement	270–1,350
Concrete	270–2,160
Brick	540–5,405
Adobe	540–5,405
Granite	2,700–5,405
Tuff	2,700–16,216
Gypsum (natural)	135–540
Concrete with alum shale	8,101–67,567
Gypsum (plasterboard)	13,510–54,050
Fly ash	1,350–8,108

Source: Modified from Stranden 1987.

these factors, so, too, is radon in building materials. Obviously, the lower the permeability, the less the effect of the other factors.

The rate at which radon is exhaled from building materials is described in terms of the emanation coefficient (see Stranden 1987). Table 5.7 shows some typical emanation coefficients for common materials. The highest emanation coefficients are noted for concrete. This is due in part to the fact that by inclusion of gravel into cement to make concrete, concrete possesses both higher porosity and permeability than cement. Emanation coefficients from brick, fly ash, gypsum, and adobe (not shown in table 5.7) yield low emanation coefficients. Stranden (1987) summarizes studies of radon exhalation rates for natural materials and man-made materials and notes that Alum Shale concretes yield the highest values, followed by concretes and brick. For walls or slabs of building materials, Alum Shale concrete yields the highest values, followed by by-product gypsum, concrete, and brick. He also notes that for homes in Sweden built using the Alum Shale concrete, indoor radon values from this source alone may yield values as high as 22 pCi/L air. More realistic values for other concretes (0.5 pCi/L), brick (0.14 pCi/L), and by-product gypsum (0.28 pCi/L) should be noted.

Should certain building materials be eliminated from use in homes? This is a very complex question. Even if the source materials may have high uranium and radium levels, without follow-up experiments, it is not possible to determine the actual amount of indoor radon due to these materials. The problem is further compounded by the fact that in most homes, there are several types of building materials, and all would

TABLE 5.7
^{222}Rn Emanation Coefficients for
Common Building Materials

Material	Emanation Coefficient
Concrete	0.1–0.4
Brick	0.02–0.1
Gypsum	0.03–0.2
Cement	0.02–0.05
Fly Ash	0.002–0.02

Source: Stranden 1987, who summarizes the work of eight other sources.

have to be assessed; a way to quantitatively assess the inputs from soil, air, and water radon would have to be known. In the Grand Junction, Colorado, case described earlier, the problem was easy to identify. This is not the case in most instances. Certainly, if materials are found to contain radiologically unacceptable amounts of radium and uranium, such material should either not be used for construction purposes or their use should be restricted to places where impact on public health is minimal (i.e., out of doors).

Remedial action on high radon-emanating building materials can be straightforward if it has been determined that the building materials in question are the real source of the radon. Sealants on materials can severely restrict radon emanation, and in some instances where the suspect material is used sparingly, it can be replaced. In other instances, more severe steps, such as structure razing and replacement (as in Grand Junction), may be necessary.

RADON FROM BEDROCKS

Elsewhere in this chapter, I have mentioned the variable nature of radon from different rocks. Soils developed over high uranium and/or high radium bedrock have a greater chance of higher rates of radon release than soils developed over uranium-poor rocks. Yet the problem in many instances is more complex. Consider two granites, each with moderately high uranium and radium contents. In one hypothetical case, the granite is highly fractured, but it is not fractured at all in the second case. The importance of the fracturing, of course, is that it facilitates radon release from the granite into the soil overlying it. In the unfractured granite, much of the radon release demands diffusion as a mechanism for the radon to reach the edges of grains, and then the radon travels by inter-granular transport only in a tight matrix. This is why permeability and porosity information about the rocks is so important.

Many rocks have pronounced and preferred orientation of minerals in them. This is true for many sedimentary rocks and for several meta-morphic rocks as well. If the stratified rocks are flat-lying, the easiest path for radon transport is parallel to the stratification. For these cases, the radon can move only with difficulty normal to the layering. If, however, the rocks are tilted or complexly folded vertically, then many of these preferred orientations are now normal to the ground and buildings above. In such cases, radon emanation may be fairly high. Thus, the structure of the rocks as well as characteristic permeability and porosity are important.

Studies in the United States and in other countries have not fully addressed the role of bedrock in the radon story. While uraniferous rocks such as the well-known Reading Prong in northeastern United States and Alum Shale in Sweden are known to be obvious radon generators, many other rocks may be as important.

CONCLUSIONS

The radon budget of various soils and rocks is often difficult to characterize, but it is even more difficult to quantitatively predict the release of radon from such media. Factors such as moisture content, porosity, temperature, permeability, soil structure, mineralogy, moisture flow direction, and others vary widely in nature, and all these effect how radon will behave in such media.

However, certain empirical observations do allow estimation of radon-prone release areas. Obviously, an ideal way to gather important data is for the EPA or some other agency to work with contractors in selected locations to gather badly needed soil property and radon data. By doing this, not only would a valuable data base be obtained, but also, where high radon release potential occurs, proper steps in preventing radon entry to dwellings in those areas could then be taken.

Waters from city supplies are usually not high in radon, but well waters may locally be extremely high. Only through programs of education and monitoring can the full impact of water-released radon be assessed.

The problem of radon entry into dwellings calls for an enlarged educational program on the part of the EPA, NRC, and DOE as well as an appreciation of the problem by industry and the general public. People who deal with developed real estate sales should be especially educated on the subject, as should local, state, and national officials. All too often this is not the case.

6

Radon Detection Methods

Radon (meaning primarily ^{222}Rn) is radioactive. So, too, are radon daughters. Most measuring devices for radon and its daughters take advantage of this fact. While there are numerous known types of detectors for radon measurement only a few are important enough to warrant inclusion in this chapter.

Briefly, radon can be measured by quick grab-sampling techniques and by more time-consuming "time integrated" methods. Each of the methods described below has its own unique advantages and disadvantages.

First, however, as we learned in chapter 5, many factors affect the way radon enters a structure. Consequently, if a "quick" measurement is made at time of either maximum or minimum entry, then this reading will not be representative of the radon level in the structure. Only by repetitive measurements of a "quick" nature can a good time-integrated average be obtained. This is illustrated in the hypothetical radon variation for some structures in figure 6.1. (Note: This figure is not based on actual data and is shown in a generic sense only.) It is obvious from inspection that grab samples taken over a few moments will represent one point on this complex radon-versus-time curve, and even sampling over a one to three day period can lead to serious errors. Of the methods to be described, the long sampling time using alpha detectors offers the best time-integrated average for indoor radon measurements.

FIGURE 6.1
Variation of indoor radon with time in a typical (hypothetical) USA dwelling. See text for detail.

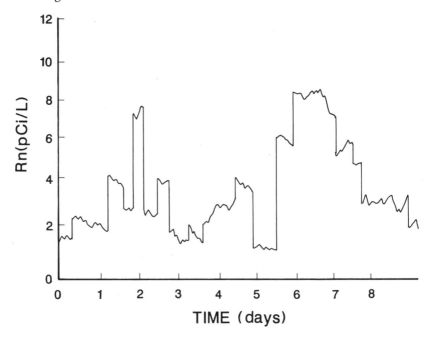

Yet, therein lies the problem. If one is interested in determining the radon level in a home and time is not a factor, then deployment of the long sampling time alpha detectors for one year (or two detectors in sequence for six months each, or four detectors over successive three month periods) is practical and reliable. But let us consider a case in which one wishes to sell a house, and the prospective buyer or the realtor wants to know the radon level now. Or perhaps, someone who is curious about indoor radon—having heard of high radon readings in the neighborhood—wants the data as soon as possible. In such cases, the grab-sampling devices and/or charcoal canister methods, both described below, may be more satisfactory, even though the radon measurement thereby obtained is a poor indication of one's average radon exposure. Still, the grab-sampling methods do have the advantage that the data are available when the sample is taken, although the charcoal canisters must rely on measurements made in a laboratory removed from the sampling site. Because each structure may be unique, each one should ideally have a radon measurement made. Since this is neither practical nor likely, radon measurements are taken in shotgun fashion in most instances. Thus, when looking over the different sets of data now available to consumers, it is imperative to know just how the data were obtained and how representative they are.

OVERVIEW OF MEASURING TECHNIQUES

Many methods are used to obtain results for radon levels. The most common will be described in this chapter, along with their advantages and disadvantages.

In the case of continuous monitoring for both radon and radon daughters, the instruments are deployed for up to thirty hours. The first few hours are for screening, equilibration, etc., and then continuous data are taken thereafter. Although these measurements are taken over a short duration, the concentrations of radon or radon daughters can be monitored on an hourly basis, the data are precise, and the data are available at the conclusion (or even during) the testing. One disadvantage is that these methods are fairly expensive (they typically cost from $100 to $500). Short-term, continuous monitoring is largely dependent on the types of services performed, and the group or individual who performs the work must be highly qualified in order to operate the equipment. Because radon concentrations vary from day to day, the data from short-term testing can be used only as an index of the radon or radon daughter level in one short period of time, which may or may not be typical.

Grab-sampling techniques for radon or radon daughters involve simple sampling; however, the samples must be taken to a laboratory for measurement. The advantages of these methods are that results are obtained quickly, the equipment is portable, several samples can be evaluated each day, and the conditions under which the samples are taken are known to the operator. The disadvantages are that this method, too, is fairly expensive, ranging from $100 to $400, the short duration of the sampling may not be representative of the radon or radon daughters in the air at other times, and the home must be carefully closed down for twelve hours before tests can be performed by a skilled operator.

Radon daughter (progeny) measuring units can be deployed and measurements can be taken over several days. The advantages of these methods are that they directly measure the conditions of the radon daughters, the measurement period is fairly short, the detector assembly (and often the whole unit) can be sent for analysis by mail, and measurement techniques in this area are very well known. The disadvantages are that these are also fairly expensive, ranging from $80 to $300 (or more), some units are heavy and awkward to move, some units can only be deployed and collected by trained personnel, and finally, sampling errors may be larger for the radon daughters because of the way they are affixed to surfaces and the way they react.

Activated charcoal canisters are very popular because they are inexpensive and provide quick data. They cost about $10, but they are often marked many times above this price for delivery, pickup, consultation, etc. The advantages, in addition to low cost, are that they can be distributed by mail, no special skills are needed to deploy or collect them, they are totally passive and need no external power, and in most cases they can yield precise data for the period of deployment. On the negative side, these detectors are sensitive to changes in household temperature and moisture, so they are limited to short-duration testing.

Alpha track detectors are also popular and have advantages that their normal cost is about $25, no special skills are needed for installation, they can be distributed by mail, and they are entirely passive and need no external power. They are not sensitive to changes in household temperature and moisture, so they can be deployed for over a year if desired. The only disadvantage is that they must be deployed for at least three months to obtain good data.

Recently, a passive monitoring method using an electrostatically charged disc that measures radon indirectly has been developed. It shows considerable promise.

CLOSED-HOUSE CONDITIONS

Prior to any radon or radon daughter measurements, the EPA says that care must be taken to ensure that the house meets closed conditions for at least twelve hours prior to the testing. This is an attempt to provide the homeowner with a "worst case" measurement that will not be adversely affected by excess ventilation, fans, etc. Normal opening and closing of doors is permitted, however. In northern climates, testing for the short-duration measurements is usually made in the winter months because in such climates the houses are usually more sealed overall than during the summer months. Hence, if the radon level is elevated, it is more likely to show up in the winter period than in the summer. To make certain that closed-house conditions are best realized, the EPA recommends that for at least twelve hours prior to the test, windows be closed, that window fans and other ventilation units not be in operation, and that routine opening and closing of doors be kept to a minimum. In addition, testing should not be carried out during periods of high winds or storms.

Detectors in closed houses should be placed away from drafts (i.e., due to vents for heating or cooling, fireplaces, windows, doors, etc.) and not too close to outside walls. The EPA also recommends installation at a height at least 50 centimeters above the floor.

Radon Sampling

Grab-Sampling Method.   Methods for quick, grab-sampling of radon in air are carried out using a continuous radon monitor. In this method, air is pumped through a filter into a scintillation cell. The filter prevents dust and radon daughters in the air from entering the cell. As radon decays, the radon decay alpha particles strike the zinc sulfide (ZnS) phosphor coating in the cell, causing scintillations. A photomultiplier tube connected to a window in the scintillation cell detects the scintillations, and electric signals are generated. These may be stored or, more often, processed so that the data are available on the spot. In order to get reliable radon data from the scintillation counts, the continuous monitor must be very carefully calibrated. The monitors use either an input of continuous flow or repetitive fillings of the cell.

The sample time given in table 6.1 of five minutes assumes that only alpha particles from ^{222}Rn will be used. For better results, the analyses can be performed after a wait of three hours, at which time sufficient ^{218}Po and ^{214}Po have accumulated (see chapter 3 for decay chain half-lives, etc.), with a resultant greater alpha decay rate. Again, the choice is the consumer's. After five minutes, quick if somewhat imprecise, data is

TABLE 6.1

Sampling Times for Different Radon and/or Radon Daughter
Measurement Techniques

Instrument	Sampling Times
Grab Rn	5 minutes
Grab WL[a]	5 minutes
Continuous Rn monitor	6 hours minimum; 24 hours or longer preferred
Continuous WL monitor	6 hours minimum; 24 hours or longer preferred
Radon progeny integrated sampling unit	100 hours minimum; 7 days preferred
Charcoal Canister	2 to 7 days
Alpha track detector	3 months desirable; less for some cases
E-PERM method	Days to months; can be reused a number of times

[a]WL = working level; see text for discussion.
Source: EPA 1986.

available. Subsequently, after several hours, more precise data can be achieved. Nevertheless the data may not reflect the true average of the structure (see fig. 6.1).

Continuous Radon Measurement. The apparatus for continuous radon measurement is identical to that for the radon grab samples, but here the intent is to obtain instant radon data more precisely and accurately. The air is pumped through a filter into the scintillation cell, and data are taken after several hours. Data taken very early will not be reliable because the air in the unit has not equilibrated properly.

Starting at about three hours, data are taken more or less continuously for a period of 24 hours. If more reliable data are desired, then radon measurements may be taken for longer periods of time. As in the grab-sampling method, the same criteria of careful calibration of the scintillation cells in known radon environments, of pumps, and of other parts of the system, are necessary. However, even when the equipment is run for 24 hours, the data may not be representative of the radon level in the structure (see fig. 6.1).

Grab Working-Level Method.   In this method, the radon daughters are measured and data are reported in units of working levels (WL). To calculate the radon in the same air sampled, it is necessary to determine or assume the degree of equilibrium between radon and the radon daughters. In the introduction I pointed out that there commonly exists disequilibrium, and that 1 WL may be equal to 100 to 200 pCi/L. This must be kept in mind when obtaining radon data from WL sampling (by both grab and continuous methods). Yet the WL data are important because it is the radon daughters, and especially ^{214}Po and ^{218}Po, that are suspected to be the prime agents for promoting lung cancer.

In the grab WL method, the radon daughters are collected from a large known volume of air on a filter. The activity of the radioactive daughters is then counted, and data are obtained by comparing this information to carefully calibrated standards. There are several methods for these determinations now available, the two most commonly used methods being the Kusnetz procedure and the Tsivoglou procedure.

In the Kusnetz method, the radioactive radon daughters are collected from about 100 liters of air over a sampling period of five minutes. After 40 minutes and before 90 minutes, the activity of the radon daughters is determined. Gross alpha counts are determined by a scintillation counter, and, as a function of the counter, converted to disintegrations. The disintegrations from this known volume of air are then converted to working levels using a series of Kusnetz factors (i.e., these factors will vary as a function of time, etc.) tabulated by the EPA (1986). The data for the Kusnetz method are obtained by the following equation:

$$WL = \frac{C}{K_{(t)} \, V \, E}$$

where C = sample counts per minute minus background counts per minute, $K_{(t)}$ = Kusnetz factor (i.e., function of time of counting), V = total air for sample in liters (i.e., flow rate [L/m] × time [m]), and E = counter efficiency in cpm/dpm (counts per minute/disintegrations per minute.

In the Tsivoglou method, the same apparatus is used, but the radon daughters on the filter are counted at least three times: between 2 to 5 minutes, 6 to 20 minutes, and 21 to 30 minutes. This allows determination of concentrations of the three important radioactive radon daughters, ^{218}Po, ^{214}Pb, and ^{214}Po (see fig. 3.1 for half-lives). The concentrations of ^{218}Po(C_2), ^{214}Pb(C_3), and ^{214}Po(C_4) are determined from the gross alpha counting rates (G_1, G_2, G_3) taken at the three time intervals, respectively:

$$C_2 = \frac{1}{FE} \ (0.16746 \ G_1 - 0.0813 \ G_2 + 0.0769 \ G_3 - 0.0566 \ R)$$

$$C_3 = \frac{1}{FE} \ (0.00184 \ G_1 - 0.00209 \ G_2 + 0.0494 \ G_3 - 0.1575 \ R)$$

$$C_4 = \frac{1}{FE} \ (-0.-0235 \ G_1 + 0.0337 \ G_2 - 0.0382 \ G_3 - 0.0576 \ R)$$

where the constants are based on the half-lives involved (see fig. 3.1), and F = sampling flow rate in liters per minute (Lpm), E = counter efficiency in counts per minute/disintegrations per minute (cpm/dpm), and R = background counting rate in cpm.

The WL is obtained from:

$$WL = (1.028 \times 10^{-3} C_2 + 5.07 \times 10^{-3} C_3 + 3.728 \times 10^{-3} C_4)$$

Continuous WL Sampling. In this method, as in the grab sample method, particulate radioactive radon daughters are captured on a filter, but here the flow rate is only 0.1 to 1.0 liters per minute. The radioactive alpha emitters caught on the filter are measured by a variety of alpha-activity detectors, all with capability to measure ^{218}Po and ^{214}Po alpha particles with energies between 2 and 8 MeV. Only the alpha particles are significant under these conditions. The apparatus is carefully calibrated in terms of time, radon levels, and counting statistics and is capable of providing hourly and total (integrated) data. The measurements are taken over at least 24 hours. As in the case of the grab samples, the calculation of radon level from the WL data requires assumptions about equilibrium between radon and radon daughters.

Radon Progeny Integrated Sampling Units for WL Determinations. In this method, air is continuously pumped through a filter and a detection unit. The air is pumped at 0.1 to 4 liters per minute. The detection unit consists of two or more thermoluminescent dosimeters (TLDs), commonly lithium fluoride (LiF), that have been previously calibrated. One TLD measures the radiation from the radon daughters, and another the background beta and gamma radiation. The measurements cannot be made on site, but must be made in a laboratory. The thermoluminescent material responds to interaction with radiation by forming whole-electron defects that, when heated, return to the original state and emit light. The amount of this thermoluminescence is a function of how much radiation it has received over a given time. The alpha-TLD is placed adjacent to the face of the filter so as to measure the alpha activity from the radon daughters entrapped there. The second TLD is placed behind a stainless steel shield to measure the beta and gamma activity. An advantage of the method is that the TLD can be cleaned and annealed for reuse by simple heating in an oven. Calibration of pumps, TLDs, and counting apparatus is an ongoing necessity. A seven-day sampling period is recommended by the EPA (1986), although times for sampling as short as 100 hours are possible. Control dosimeters are provided in each detection unit to check for nonradiation induced thermoluminescence and extraneous alpha radiation. Problems with this technique have been discussed by Nazaroff (1987).

Charcoal Detection.   Activated charcoal canisters are very popular with many radon measurement firms and individuals. These consist of tightly sealed canisters of activated charcoal that will allow air to filter into them when the caps are removed. The charcoal attracts the radon, and radon daughters start to accumulate due to radioactive decay.

The canisters are typically deployed for one to seven days, then the sealing caps are replaced and the canisters immediately sent to a laboratory for analysis. Because of the short half-life of ^{222}Rn (3.8 days), it is important to have the analysis done very quickly after the deployment period so that enough radon daughter product is available for measurement. Using standard counting equipment, the detection is made by counting gamma rays of energies from 0.25 to 0.61 MeV. The charcoal is water sensitive, and a correction must be applied for water. The amount of water is calculated by comparing weights of the canister plus charcoal before and after the deployment with the difference attributed to water.

While these canisters give reasonable data (see Cohen and Cohen 1983; George 1984), the sampling times are sufficiently short so that a time-integrated average for radon is not assured (fig. 6.1). Nevertheless, these canisters are suitable for fairly quick turnaround time and thus are used to obtain radon data when speed of analysis is important. If time permits, repeated analyses on a monthly basis is desirable.

Alpha Track Detector Measurements. The use of alpha track detectors is widespread for routine radon measurements. This method takes advantage of the fact that certain emulsions (special organic compounds, in particular) are damaged by impact of alpha particles, as is shown in figure 6.2. When the emulsion containing the damaged area—the so-called track—is leached, the track becomes visible. The track density is dependent on time, amount of radon, and volume of air. The most commonly used detectors require a filter that prevents radon daughters from entering the detector and, often, a membrane to prevent thoron (^{220}Rn) from entering as well. The plastic is thus bombarded only by the ^{222}Rn in the cup space plus those alpha emitting radon daughters (^{218}Po and ^{214}Po) formed therein. If left for sufficient time, the air in the detector cup is in equilibrium with the surrounding air. This method has the advantage that even if the radon levels in the air should change, the tracks will record the time-integrated variations such that a meaningful average radon level can be determined. In figure 6.1 it is noted that testing periods of under a month may be suspect; hence, if an alpha-detecting device is deployed for, say, three months (or longer if possible), then realistic data can be obtained.

While this book is concerned with indoor radon, it must be pointed out that these alpha detectors are also widely used for soil, air, and water measurements as well. Other applications are given in chapter 11.

The Electret-Passive Environment Radon Monitor. Recently, a novel way of measuring radon concentrations has been developed: the electret-passive environmental radon monitor, or E-PERM. In this system, which costs about $10 to $25, an electrostatically charged plastic disc, called an electret, is mounted in a small canister isolated from particulates by a filter. Radon passes through the filter and induces minor negative charge in the air near the positively charged electric. The negative ions are attracted to the electret surface and a voltage is imparted. This measured voltage is proportional to the radon content of the air and can be corrected for length of exposure, background charge, and other factors. The method can be deployed for short or long periods, depending on the

FIGURE 6.2

Terradex Type SF indoor radon detector. The detector contains an alpha-sensitive emulsion inside protected by a filter to keep out particulates and thoron. See text for detail. Figure is courtesy of Terradex Corporation, a subsidiary of Tech-Ops Corporation.

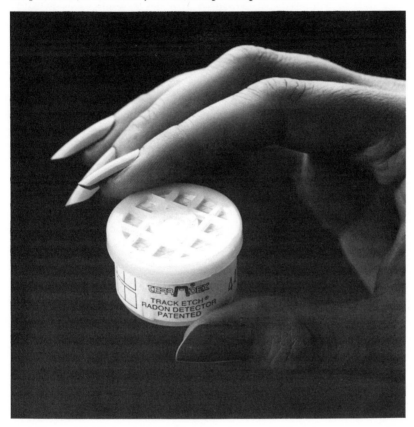

electret used. The electret can also be reused a number of times until the charged surface has been effectively neutralized, perhaps after some ten uses. Further, the electret is not apparently sensitive to humidity (as the charcoal canisters are) and may give results equal to those obtained from alpha track measurements over long-term deployment situations. This method has been described by Kotrappa and his coworkers (see references for this chapter).

CONCLUSIONS

There are many radon measuring techniques available at the present time; some are better than others. Grab-sampling methods and the use of charcoal canisters are useful if swiftness in obtaining data is more important than the reliability of the data. Figure 6.1 illustrates just how extreme radon variability can be over a short period of time. When quality data, representative of a time-integrated approach, are required, alpha track detectors or long-term continuous monitoring must be employed.

It must be emphasized that any single radon analysis will provide one datum only, and it is time-consuming and expensive to properly determine a dwelling's radon budget. Nevertheless, if any one radon reading is above 4 pCi/L, I recommend follow-up analyses to verify the number and inspect other rooms in the dwelling. Remember, it is your health that is at stake!

7

Radon Studies in the United States

Indoor radon is now a new buzz phrase in several parts of the United States and is growing rapidly in many other places. With the recognition of indoor radon as a potential health threat and the concern of citizens, the demand for radon data has increased. A few years ago, there were perhaps a dozen firms nationwide that handled routine radon surveys; now there are several thousand. Further, it is widely reported that many of these radon monitoring firms are not reliable—some, perhaps, are blatantly dishonest.

Let the buyer beware. The EPA provides a list of firms that have had their radon-monitoring devices approved by EPA laboratories, and consumers are urged to consult this list before getting radon measurements made. It is also reported that there are firms combining questionable radon measurements with even more questionable remedial work (see chapters 8 and 10). Again, consumers are advised to carefully investigate the radon-monitoring services first, and only after being convinced that there is a potential problem should they proceed with any kind of remedial work. Consumers should choose contractors with demonstrated knowledge of the radon problem and experience (preferably) in working with remedial procedures. And finally, consumers should be aware of prices: Radon detectors cost from about $10 to $25 (activated charcoal and alpha track, respectively), yet many firms charge $100 and more.

RADON MEASUREMENTS AND CONCENTRATIONS IN THE UNITED STATES

There have been many measurements of indoor radon in the United States. The firm Terradex Corporation, a subsidiary of Tech-Ops, in Illinois, has an extensive data base. Its data for some 60,000 measurements from across the United States yields an arithmetic mean of over 4 pCi/L for the sample, with about 20 percent above 4 pCi/L. The Terradex report has been criticized by Nero (see Nazaroff and Nero 1987) who points out that many measurements were from basements, that replicate data were not obtained, and that radon-prone areas may have been over-sampled. Nevertheless, the Terradex data are an extremely large sample; the 60,000 measurements represent (approximately) 30,000 homes, or about 0.05 of all single family dwellings in the United States. The data are given in table 7.1, modified slightly from the original source.

The data are heavily skewed toward certain states, though. Pennsyl-

TABLE 7.1
Data Base Summary by State

State	No. of Meas.	pCi/L Median	pCi/L GM[a]	pCi/L AM[b]	Percent Greater Than 4 pCi/L	Percent Greater Than 8 pCi/L	Percent Greater Than 20 pCi/L	Maximum Reading pCi/L
AK	59	1.54	1.54	3.60	32.2	16.9	.0	18.21
AL	535	1.33	1.49	2.85	17.0	7.7	2.1	49.94
AR	4	3.37	3.97	4.29	50.0	.0	.0	7.19
AZ	92	.69	.96	6.65	21.7	15.2	7.6	89.80
CA	688	.78	.76	1.40	4.5	1.5	.4	67.88
CO	266	3.61	3.89	11.98	45.9	16.5	6.8	691.31
CT	158	1.93	3.94	4.04	19.6	5.7	2.5	152.38
DC	48	.78	1.11	4.38	14.6	8.3	8.3	89.34
DE	37	1.12	1.37	2.07	16.2	2.7	.0	10.55
FL	240	1.23	1.40	5.29	16.7	8.3	6.3	90.47
GA	73	1.05	1.25	2.04	13.7	22.7	.0	19.25
HI	1	.36	.36	.36	.0	.0	.0	.36
ID	787	3.02	3.24	6.13	41.0	12.0	6.0	66.62

State	No. of Meas.	pCi/L			Percent Greater Than			Maximum Reading pCi/L
		Median	GM[a]	AM[b]	4 pCi/L	8 pCi/L	20 pCi/L	
IL	177	1.73	2.10	9.10	23.7	6.8	5.1	196.00
IN	183	2.79	2.82	5.94	35.5	14.2	3.8	167.39
IA	67	2.26	2.23	3.61	32.8	6.0	1.5	25.35
KS	13	1.97	2.62	4.51	23.1	15.4	7.7	24.79
KY	102	1.62	1.80	3.72	25.5	17.6	2.0	34.55
LA	25	.61	.64	.95	4.0	.0	.0	4.68
ME	1,524	1.75	1.95	11.99	24.3	10.8	4.9	4,354.47
MD	196	2.23	2.33	4.96	25.0	13.8	3.1	97.02
MA	213	1.55	1.46	2.43	16.9	3.8	.5	31.00
MI	92	1.91	1.87	3.19	22.8	9.8	2.2	20.67
MN	249	2.52	2.92	5.76	36.5	17.7	3.6	142.48
MS	15	1.22	1.15	1.35	.0	.0	.0	2.48
MO	60	.93	.94	1.55	10.0	.0	.0	7.63
MT	321	2.95	2.74	4.33	33.0	12.8	2.5	29.58
NE	15	2.64	2.95	4.55	33.3	13.3	.0	17.12
NV	311	1.51	1.70	4.45	14.1	7.7	5.8	67.45
NH	95	3.06	3.19	6.41	40.0	21.1	8.4	72.38
NJ	4,475	1.96	2.16	6.86	26.3	12.3	4.4	1,234.55
NM	580	2.03	2.17	4.30	26.2	9.3	2.6	213.84
NY	4,906	1.04	1.09	2.26	11.7	4.6	1.1	107.92
NC	92	1.38	1.25	2.35	13.0	4.3	1.1	30.09
ND	49	2.58	3.10	4.74	28.6	16.3	4.1	33.06
OH	433	3.00	2.91	5.09	38.8	15.9	3.2	63.21
OK	16	1.49	1.72	3.07	18.7	12.5	.0	13.61
OR	5,835	.81	.81	1.33	5.0	1.2	.1	67.25
PA	22,312	4.89	5.14	14.14	56.8	33.2	12.1	3,125.39
RI	20	1.85	2.84	12.38	30.0	25.0	10.0	153.18
SC	26	.84	.83	1.10	33.8	.0	.0	5.96
SD	605	2.27	2.45	6.65	28.3	13.1	6.0	128.74
TN	502	1.92	2.08	3.20	26.3	9.6	.4	23.53
TX	464	.60	.57	.79	.6	.2	.0	14.63
UT	29	2.07	1.98	2.77	17.2	6.9	.0	9.67
VT	71	1.69	1.95	3.32	23.9	9.9	1.4	21.85
VA	488	1.05	1.04	2.20	12.9	4.9	.4	45.91
WA	11,711	.54	.55	1.05	3.8	1.3	.2	92.41
WV	15	1.79	1.74	6.44	20.0	13.3	6.7	65.27
WI	225	1.85	1.68	2.57	19.6	4.4	.0	12.77
WY	135	3.56	3.82	12.73	48.1	37.8	20.0	111.50

[a] Geometric mean.   [b] Arithmetic mean.
Source: Alter and Oswald 1988.

vania is represented by 22,312 measurements, Washington by 11,711, Oregon by 5,835, New York by 4,906, New Jersey by 4,475, and Maine by 1,514. The heavy sampling in these states reflects the concerns over indoor radon. In Pennsylvania, the Reading Prong is an identified source of uranium (see discussion of the Watras case in the introduction), and a large effort has been undertaken in this state to identify high-radon dwellings. Table 7.1 shows that 56.8 percent of all measurements fall above 4 pCi/L and 33.2 percent above 8 pCi/L—staggeringly high figures. The geometric mean for the Pennsylvania samples is 5.14 pCi/L. In Maine, another area with known high-uranium granites, the mean for 1,514 samples is 1.95 pCi/L, but 24.3 and 10.8 percent fall above 4 and 8 pCi/L, respectively. For Washington, uranium mineralization is common in the Spokane River Valley, and a high incidence of measurements is in this general area. In the Spokane River Valley of Washington and Idaho, 46 homes yielded a mean of 13.3 pCi/L. The geometric mean is only 0.54 pCi/L for Washington, however, and only 3.8 and 1.3 percent of all measurements fall above 4 and 8 pCi/L, respectively. In Oregon, the geometric mean is 0.81 pCi/L, and 5.0 and 1.3 percent of the data are above 4 and 8 pCi/L, respectively. Overall, the northwestern states (Washington and Oregon) show fairly low indoor radon values.

Concern over indoor radon at elevated levels has led to many data being taken in the New Jersey and New York areas. For New Jersey, the geometric mean is 2.16 pCi/L; 26.3 and 12.3 percent of the 4,475 homes fall above 4 and 8 pCi/L respectively. In the Clinton, New Jersey area, studies of 103 homes indicate a geometric mean of 100 pCi/L. It has been pointed out that, since many of the homes in this study were slab-on-grade, even without a basement, the indoor radon values were high due to the underlying high uranium- and radon-baring rocks and soil. This, in turn, shows the importance of sealing even slab-on-grade basal structures to prevent radon entry (see chapter 2 for more detail).

For the 4,906 measurements from New York, the geometric mean is 1.06 pCi/L; 11.7 and 4.6 percent of the total data fall above 4 and 8 pCi/L, respectively. About 35 percent of the Terradex data are from basements. In table 7.2 we see the breakdown for the basement versus nonbasement readings. In general, the basement readings are about twice the value for the nonbasement areas. This fact suggests an empirical way of estimating a risk factor for indoor radon by multiplying the arithmetic mean (see table 7.1) by the state's population; it can be argued that this

TABLE 7.2
Basement Versus Non-Basement Radon Statistics

Region	No. of Meas.	Median	GM	AM	Percent over 4pCi/L
United States					
Basements	9,258	4.67	4.99	14.05	55.5
Nonbasements	11,814	1.53	1.72	7.07	23.0
Ratio[a]	0.78	3.05	2.90	1.99	2.41
Northeast					
Basements	8,512	4.87	5.24	14.76	57.0
Nonbasements	7,821	1.72	1.97	9.14	26.9
Ratio	1.09	2.83	2.66	1.61	2.12
Midwest					
Basements	129	3.62	3.58	6.74	48.1
Nonbasements	282	1.93	1.93	2.90	21.3
Ratio	0.46	1.88	1.855	2.29	2.26
Northwest					
Basements	237	1.70	1.84	4.10	22.4
Nonbasements	1,454	.79	.80	1.74	8.6
Ratio	0.16	2.15	2.30	2.36	2.60
Southeast					
Basements	80	3.52	3.39	5.07	47.5
Nonbasements	690	1.49	1.57	3.55	18.6
Ratio	0.12	2.36	2.16	1.43	2.55
Mountain states					
Basements	66	4.65	5.61	12.20	60.6
Nonbasements	683	2.13	2.16	3.93	25.9
Ratio	0.09	2.18	2.60	3.10	2.34

[a] Ratio basements/nonbasements.
Source: Alter and Oswald 1988.

product may be proportional to incidences of lung cancer, although this hypothesis has not been tested.

Nero (1987) and his coworkers argue that the distribution of indoor radon values in the United States is lognormal and that the geometric mean for the United States is on the order of 0.9 pCi/L (or an arithmetic mean of 1.5 pCi/L), with a long tail of high radon values for 1 to 3 percent of homes in the United States having values above 8 pCi/L. They base their conclusions on an assessment of 38 sets of data, 19 of which they assume are more reliable for defining radon distribution in dwellings in the United States. While the total number of homes measured, 522 for the 19 data sets chosen as best examples for this study, is not large, the geographic coverage is large enough that the authors argue for acceptance of their figures for indoor radon distribution. Figure 7.1 is a redrawn version of their figure 1.1.

As I mentioned in the introduction, a study of 74,000 homes in the greater Washington, D.C., area, including parts of Maryland and Virginia, revealed that about 25,000 homes (one-third) had indoor radon levels over 4 pCi/L. This information, obtained by the use of charcoal detectors, is valuable only for screening purposes. Nevertheless, the magnitude of the number of homes over 4 pCi/L (table 7.3) is truly staggering, and it further points to the public health issue of radon.

Meanwhile, the radon problem is far from being solved. While it has been pointed out that the lognormal distribution may mask many high radon values because of the small size of samples and other factors, it

FIGURE 7.1
Probability distribution of ^{222}Rn in USA dwellings according to Nero (1986). Fig. 8-1 is based on 522 data. See text for detail.

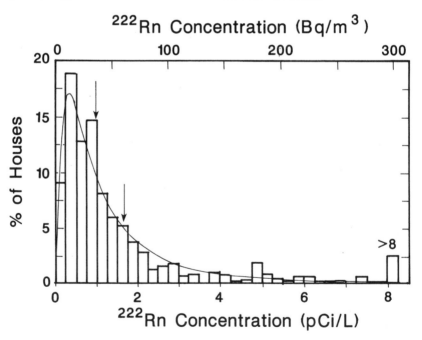

TABLE 7.3

Population Weighted Arithmetic Mean for 22 States with Greater than 200 Measurements

State	Population (millions)	No. of Meas.	AM (pCi/L)	AM × Population
PA	11.9	22,312	14.1	168.3
FL	11.5	240	5.33	60.8
OH	10.6	433	5.1	54.0
NJ	7.1	4,475	6.9	48.7
NY	17.9	4,906	2.3	40.5
CO	3.0	266	12.0	35.9
CA	24.1	688	1.4	33.3
MN	4.0	249	5.8	22.8
ME	1.4	1,524	12.0	16.8
TN	4.6	502	3.2	14.7
MA	5.5	213	2.4	13.4
WI	4.8	225	2.6	12.3
VA	5.3	488	2.2	11.6
TX	14.5	464	0.8	11.5
AL	3.7	535	2.9	10.5
NM	1.4	580	4.3	6.0
ID	0.9	787	6.1	5.5
SD	0.7	605	6.7	4.7
WA	4.4	11,711	1.1	4.6
NV	0.9	311	4.5	4.0
OR	2.8	5,835	1.3	3.7
MT	0.4	321	4.3	1.7
Total	141.4	57,670	4.14[a]	585.3

[a] Population-weighted mean = 4.14 pCi/L.
Source: Alter and Oswald 1988.

can be argued that the use of the arithmetic mean of the Terradex report of over 4 pCi/L for homes in the United States is unrealistically high. Clearly, there is a need for the EPA, the DOE, and the NRC to come to grips with this problem and to try to identify the best means of obtaining a representative sampling across the United States for predictive health reasons. At present, measurements are largely in response to public or private request; there is no uniformity of types of monitoring devices employed, no criteria for where samples are to be placed, assessment of the dwellings for uniformity of insulation, venting, building materials, water supply, and other factors, and no systematic year-round sampling.

For the last few years, Dr. Douglas Mose of George Mason University has been conducting detailed indoor radon measurements in Fairfax County, Virginia, and Montgomery County, Maryland. His approach is to have homeowners pay for alpha detectors (see chapter 8) quarterly, so that data are compiled for the four seasons and he can thus assess the effects of closed versus more open home conditions. He finds (written communication, April 1988) that the annual mean indoor radon level is within a factor of two of the high and low values. His studies, based on about a thousand dwellings, have yielded very important information. He has shown, for example, that the fall and spring mean indoor radon values are, statistically, the best indicators of the annual values, with summer values being lower and winter values being higher than those for fall and spring. Yet all are within a factor of two of the annual mean values.

Many of Mose's data are from basements, but some data are from nonbasement areas. Part of his data are reproduced as tables 7.4 and 7.5, in which it is noted that while the means for the basement readings

TABLE 7.4
Summary of Indoor Radon from Participating Homes in Fairfax County, Virginia

Measurement Season	Median Rn in pCi/L	% Over 4 pCi/L	% Over 10 pCi/L	% Over 20 pCi/L	Number of Homes	Data from Basements
Winter 1986–1987 and 1987–1988	3.7 pCi/L	44%	11%	2%	767	86%
Spring 1987 and 1988	2.7 pCi/L	30%	5%	1%	744	83%
Summer 1987	2.3 pCi/L	21%	2%	0%	700	82%
Fall 1987	3.0 pCi/L	32%	3%	0%	750	83%
Basement Only Data						
Winter	3.9 pCi/L	47%	9%	2%	272	—
Spring	2.9 pCi/L	32%	6%	2%	457	—
Summer	2.4 pCi/L	20%	2%	0%	713	—
Fall	2.8 pCi/L	24%	4%	0%	659	—

Source: Mose 1988 (written communication).

TABLE 7.5
Summary of Indoor Radon from Homes in Montgomery County, Maryland

Measurement Season	Median Rn in pCi/L	% Over 4 pCi/L	% Over 10 pCi/L	% Over 20 pCi/L	Number of Homes	Data from Basements
Winter 1986–1987 and 1987–1988	3.7 pCi/L	48%	16%	4%	229	83%
Spring 1987 and 1988	3.2 pCi/L	38%	10%	2%	160	79%
Summer 1987	2.5 pCi/L	25%	3%	1%	214	87%
Fall 1987	3.1 pCi/L	36%	6%	1%	216	82%
Basement Only Data						
Winter	4.4 pCi/L	52%	15%	3%	67	—
Spring	3.6 pCi/L	43%	10%	1%	67	—
Summer	2.8 pCi/L	32%	3%	1%	273	—
Fall	3.3 pCi/L	24%	7%	1%	231	—

Source: Mose 1988 (written communication).

are always higher than those for the nonbasement readings, the difference is usually on the order of (basement higher) + 0.2 to + 0.6 pCi/L. In other words, a reasonable estimate of the nonbasement living areas can be made from the basement data. Mose also notes that basements with cinder block walls are more prone to high radon than basements with cement walls and that the "basement effect" is commonly observed on the first floor above the basement. He is at present undertaking to test the correlation of the indoor radon values with the geology of the two counties, soil infiltration parameters, and other factors. I gratefully acknowledge Mose's permission to use his data here.

Mose's data are also important for the reason that his studies cover entire years. His mean indoor radon level of 4 pCi/L is significantly higher than that predicted by Nero and his coworkers, whose model is based on a lognormal distribution. Basing his study on about 100 sites, Mose correlates soil radon with soil infiltration (assumed to be proportionate to permeability). He notes (written communication, 1988) that in terms of soil radon versus soil infiltration data, he can predict the general indoor radon range for the area. For 100 homes in his low infiltration, low soil-radon group, his estimates were correct in 83 percent of the cases, and he was 70 percent correct in the medium-risk soil-radon-soil infiltration group for ten homes (i.e., indoor radon levels of 5 to 15 pCi/L). Although still in an early stage, his data are impressive. Figure 7.2 is reproduced from Mose's April 25, 1988, newsletter.

Table 7.1 shows that most indoor radon readings are from areas other than the arid southwestern states. The combined states of Nevada, Utah, Arizona, and New Mexico, provide only 1,012, or 1.6 percent, of the total 61,014 (Alter and Oswald 1987; Terradex). Yet there are moderate to large metropolitan cities in these states: Las Vegas, Salt Lake City, Phoenix, Tucson, Albuquerque, Denver, and El Paso could also have been included here.

RADON STUDIES IN ALBUQUERQUE, NEW MEXICO

I have been researching the problem of indoor radon in arid to semi-arid cities for some years and have found that soil uranium is not routinely flushed from the soils, as it is in more humid areas. The soils in arid to semi-arid lands are usually quite immature, and, especially in places with uranium-rich granitic or other uranium-rich detritus, soil uranium-radium-radon may be quite high.

In the winter of 1983–1984, I conducted a pilot study of soil and indoor radon in the Albuquerque area. My findings showed that soil

FIGURE 7.2
Interpretative chart for relationship between soil radon, soil permeability and indoor radon. Fig. 7-1 is redrawn with permission from D. Mose (written permission, 1988).

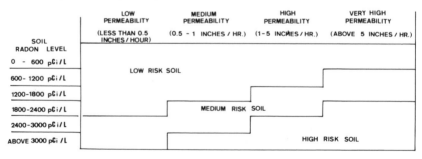

radon increased from the Rio Grande east toward the Sandia Mountains (see fig. 7.2) and that eight of the 15 dwellings tested for indoor radon yielded readings above 4 pCi/L.

The Albuquerque area consists of a down-faulted and in-filled series of sediments cut by the Rio Grande and bordered on the east by the rocks of the Sandia Mountains (see fig. 7.2). I have also studied the uranium budget of the Sandia Mountains, and some of my findings are reported in table 7.1, where it is noted that both the Sandia granite, which makes up the greater part of the mountains, and the overlying Madera Formation (limestone; Permian) contain above-average uranium compared to average crustal abundance, average granite, and average shale (note that data for uranium in an average limestone are not well-defined). In addition, the soils in the Albuquerque area are extremely immature, as is typical of most areas in the arid southwestern United States. As such, they have not been leached of their uranium (i.e., compared to many very mature soils from which uranium has been leached, with the result that their radon flux is very low). The radon flux from the Albuquerque area soils is thus controlled in large part by the granitic detritus and to some degree by limestone detritus, and the radon is in approximate equilibrium with soil radium and uranium. This, in turn, indicates that soil radon and the radon flux should be proportionate to uranium content, a hypothesis that may make it easier to pinpoint areas of possible high radon flux in other parts of the arid southwest from data taken in the mid- to late-1970s by several national laboratories and the National Uranium Resource Evaluation project (NURE).

My data for the winter 1986–1987 studies are given in the references (corresponding with this chapter) along with comparative data for winter 1983–1984 and summer 1986 (see Brookins 1988). The 1983–1984 sample locations are given in figure 7.3.

For an area such as Albuquerque, with a population of roughly 425,000, the proper sampling is problematic. I wished to obtain a wide-spread but usable data base; therefore, I advertised in the local news media the availability of free radon monitoring of homes for the winter 1986–1987 season and solicited telephone response. Prior to this, I had selected 20 homes around the city for a comparative study in the summer 1986 to winter 1986–1987. Out of 275 responses, 180 homes were selected based on geographic distribution and, in some cases, special building types and other criteria. The station locations are plotted in

FIGURE 7.3
Map of the Albuquerque, New Mexico area showing soil radon values (values in pCi/L in parentheses) for fifteen stations (dots).

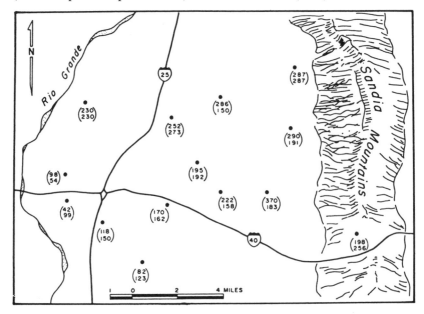

figure 7.4. It is noted that the distribution of samples, while not complete, is nevertheless very widespread and representative of the Albuquerque area.

The radon detectors for the 180 dwellings selected were deployed in October 1986 and collected at the end of February 1987. This period covers five of the six months in the area when cold temperatures prevail. The 20 dwellings selected for the summer 1986 sampling were included in this total. The summer sampling involved deployment of radon detectors in June 1986 and collection in September 1986, a period covering roughly three months of the normal high temperature period in the area.

Data for degree of equilibrium between radon and radon daughters in the Albuquerque area are lacking, yet the fact that soil uranium-radium-radon levels are in equilibrium suggests that overall equilibrium is probable.

The Albuquerque study included the winter 1983–1984, summer 1986, and winter 1986–1987 periods. I placed most of the radon detec-

FIGURE 7.4

Map of the Albuquerque, New Mexico area showing indoor radon values for approximately 200 locations. Legend: less than 4 pCi/L = open circles and squares, 4 to 8 pCi/L = closed circles and squares, above 8 pCi/L = half filled circles and squares. See text for detail. Modified from Brookins (1988).

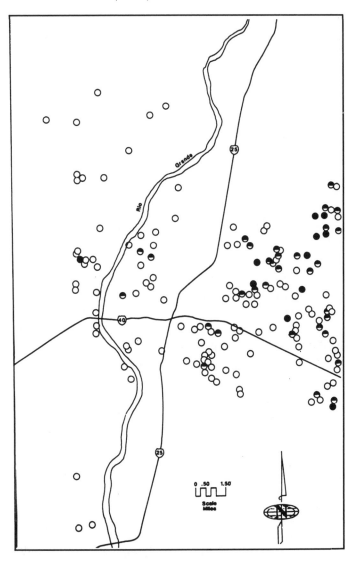

tors in areas where individuals spent at least one-third of their daily routine. This was intentional, since the study was aimed at obtaining living area values. The EPA, on the other hand, recommends choosing more isolated areas for initial screening, taking pains to keep the area somewhat closed off, a practice that may tend to maximize indoor radon effects. Because this was not done in the 1986–1987 study, it is possible that the readings on the whole may be somewhat minimized (although this remains to be demonstrated unequivocally). However, the winter 1986–1987 readings cover the period (i.e., about five months) when houses are usually closed up, so the readings are considered representative of the fall and winter months in the area.

Figure 7.4, where the results of the winter 1986–1987 study are plotted, shows that there is a higher incidence of elevated indoor radon levels as one moves toward the Sandia Mountains in the east, especially the northeasternmost part of Albuquerque. Here, most of these houses are in a fairly new, expensive neighborhood, and all dwellings are well insulated. Many are built in part over bedrock. Thus, it is not surprising that both soil (see fig. 7.3) and indoor radon levels (see fig. 7.4) are high.

The data for the winter 1986–1987 study are shown in figure 7.5. I also tabulated topics such as the year the house was built, building type (material dominant in the home construction), amount of insulation, air conditioning, whether the house does or does not have solar capacity, and the location of the detector in the dwelling. I noted a crude correlation of houses with a solar capacity and/or extra insulation with radon levels above both the geometric mean (2.7 pCi/L) and arithmetic mean (3.06 pCi/L) for the data and a lesser positive correlation with radon levels and year of construction (i.e., higher values for newer houses). This last relationship is probably more a reflection of proximity to the Sandia Mountains than to the actual year of construction, however. Many of the newer homes also have refrigerated air conditioning systems, but this is a factor only in the summertime. In most cases, the location of the detectors did not have an effect, although where basements were available for sampling, they did show somewhat elevated levels. Basements are not common in the Albuquerque area, however. Others have argued that radon entry from slab-on-grade construction may be as high as in basements.

In a comparison between the summer 1986 and winter 1986–1987 studies, 12 of 17 homes showed higher winter readings. Of the 5 show-

FIGURE 7.5

Distribution of ^{222}Rn in Albuquerque, NM dwellings based on 200 data. Compare with Fig. 8-1. See text for detail.

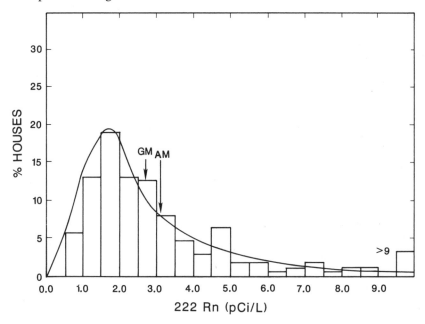

ing higher summer readings, 4 of the 5 have refrigerated air conditioning as opposed to the more common evaporative swamp coolers. To make swamp coolers efficient, windows are cracked open, thus making possible a greater flow of air circulation from outside.

Of further interest is the observation that the average difference between winter and summer readings is 1.1 pCi/L. If this figure is used to estimate summer readings for those dwellings at or above 4 pCi/L, then 24 could yield elevated levels even for the summer period, and all of the houses with elevated fall and winter readings could yield yearly averages above 4 pCi/L. This suggests that for the Albuquerque area, the percentage of dwellings with a yearly average greater than 4 pCi/L is probably close to the number with elevated winter readings.

The role of building materials cannot be assessed from my studies, but since several potentially uraniferous materials may be involved in the local building/contracting uses, this topic should be investigated further. For example, fly ash from a coal power plant is used as an additive to cement, and at times this material is uraniferous. Similarly, the pumice added to cement to make cinder block and the clay used for adobe have above average uranium content. Even if there is a measurable contribution from different building materials, however, its role must be small because the data do not show any correlation of indoor radon with such materials.

In combining the winter 1986–1987 data with the winter 1983–1984 results, I noted that 28 percent of the dwellings yield readings above 4 pCi/L. Again, while this is for the fall-winter period, this percentage (28 percent) probably holds for yearly averages as well, although the winter arithmetic mean (3.06 pCi/L) is probably higher by 0.3 to 0.5 pCi/L. Regardless, the winter values show that a large number of homes in the Albuquerque area (28 percent) possess elevated indoor radon levels compared to the United States estimated average of 10–15 percent (EPA 1986), or 7 percent as argued elsewhere by Nero (1986).

The observation (see figs. 7.3 and 7.4) that both soil and indoor radon levels increase toward the Sandia Mountains in the Albuquerque area is of broad regional interest. Although radon release in soils is a complex function of permeability, porosity, moisture content, and other factors in very immature, granitic soils such as those in this study area, there may be a close correlation of soil radon with uranium. Soil radium has been shown to be in equilibrium with radon and also possibly with uranium. Soil uranium in the greater Albuquerque area has been reported by Los Alamos National Laboratory which tabulated data for 51 stream sediments. The range in the stream sediments is from 1.8 to 14.63 ppm uranium, with an arithmetic mean of 4.70 ppm uranium ($+/-0.75$ ppm). Further, they also report a range in thorium-uranium from 2.2 to 5.6 with an average of 3.2 (± 0.6). For an arid area such as in the Albuquerque area, the stream sediment samples can be taken as representative of soil uranium as well. Thus the 4.70 ppm average uranium content is important in that this value is close to that for the Sandia Granite. The thorium-uranium ratio for the granite, however, is different from that reported for the soils. A thorium-uranium ratio for 31 granite samples of 4.3 (± 1) has been reported which is higher than the 3.2 (± 0.6) value for the data reported for soils. These results indicate that some uranium has been mobilized from the granitic rocks and fixed in the stream sediments, hence the slightly lower thorium-uranium ratios in the sediments. To test this unequivocally, samples of weathered granite (i.e., those samples most likely to have lost uranium) would have to be studied. Much of the Sandia granite has an altered reddish-to-yellowish surface, indicating oxidation of Fe^{2+} to Fe^{3+} (i.e., formation of goethite and hematite). I pointed out in chapter three, that if there is enough oxygen dissolved in waters to oxidize iron from the +2 to the +3 state, then there is more than enough oxygen in the waters to have oxidized any available uranium of charge +4 to the +6 state; and the +6 state of uranium is highly soluble. Iron minerals that contain +3 iron in them, such as hematite (Fe_2O_3) and goethite (FeO.OH), are usually fairly brightly colored; reds and yellows are common. Thus, rocks where this oxidation has occurred are easy to spot. In the Sandia granite, many of these rocks were uplifted some 60 million years ago, and they have been weathering since. Often, the exposed granite is stained yellow to brown to red, indicating the oxidation of iron +2 to iron +3 and thus implying the potential for release of some of the rocks' uranium. This released uranium may, then, also be a source of precipitated radium on grain surfaces, from which radon, when formed, can emanate.

There is an extension of these ideas that should provoke further work in the arid southwestern metropolitan areas. All these cities have common to them a set of arid conditions, exposures of rocks of silicic composition (or others of modest uranium contents), and well insulated dwellings. The data of the Natural Uranium Resource Evaluation (NURE) project can be used to indicate areas where stream sediment, and thus soil, uranium is above average (or at least elevated relative to stream sediments). These areas, then, can be pinpointed as those deserving of indoor radon screening by EPA, state agencies, or others. Often, the uranium levels are such that airborne radiometrics may be too insensitive to detect subtle anomalies, but the stream sediment surveys, involving very large numbers of samples, can allow such anomalies to be detected. Certainly, this is true for the Albuquerque area. Other high uranium areas in New Mexico have not, unfortunately, been tested for their radon contents.

I conclude that:

1. Of 175 dwellings studied in the Albuquerque, New Mexico, area during the winter of 1986–1987, 44 yield indoor radon levels greater than 4 pCi/L.
2. Of 190 dwellings, 52, or 28 percent, fall into the elevated range when

the results of the winter 1983–1984 study are combined with the results of 1986–1987.

3. Indoor radon levels increase toward the Sandia Mountains, as does soil radon.

4. Indoor radon levels may correlate positively with degree of insulation and solar capacity systems. No clear correlation is indicated for level of indoor radon with year of house construction, nor for building material, although building materials should be investigated further (see Brookins 1988).

5. Most summer 1986 indoor radon readings (i.e., 12 of 17) are lower than readings taken in fall and winter of 1986–1987. The difference averages 1.1 pC/L.

6. The geometric mean indoor radon level for the Albuquerque area for the fall-winter 1986–1987 period is 2.7 pCi/L, and the arithmetic mean is 3.06 pCi/L, for 175 samples.

7. Further study may show that the Albuquerque data are typical of dwellings built on immature, granitic soil in the arid southwestern United States.

8. Uranium has been removed only slightly from the Sandia granite, and stream sediment, and Rio Grande sediment samples show that this uranium has been locally fixed. This, in turn, supports the hypothesis of a convenient source for mobile radon in the Albuquerque area (Brookins 1988).

9. National Uranium Resource Evaluation (NURE) data may be useful when identifying other areas in the arid southwestern United States where high uranium and potentially high radon may be found.

FEDERAL PROGRAMS

The Department of Energy Program. The United States Department of Energy (DOE), with its predecessors, the Atomic Energy Commission (AEC) and the Energy Research and Development Administration (ERDA), has been involved with radon research for over forty years. The DOE program is now handled by its Office of Health and Environmental Research (OHER). The goals of the DOE/OHER program are shown in figure 7.6. Two parallel research efforts have been undertaken by DOE/OHER, the first being the characterization of the exposure environment and the second being dose-risk relationships. The former requires de-

FIGURE 7.6

Principal elements of the DOE/OHER radon research program (See text for details). From DOE.

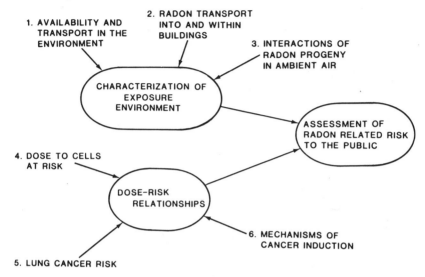

tailed study of availability and transport of radon in the environment, radon transport into and within buildings, and interactions of radon daughters in air; the latter integrates dose-to-cells at risk, lung cancer risk, and mechanisms of cancer induction. The overall goal of the program is the assessment of radon related risk to the public.

Table 7.6 shows some of the key events of the government's program in radon and related research. In 1949, standard radon measuring devices were designed with high precision in the Environmental Measurements Laboratory, which also initiated monitoring and measuring in mine atmospheres in 1951. The Argonne National Laboratory developed a radon breath analysis technique in 1952. In 1953, the University of Rochester initiated studies of inhalation effects on mice of radon daughters. By 1956, higher precision radon measuring techniques, scintillation flash counting methods, were developed at both Argonne National Laboratory and Los Alamos National Laboratory. Continued work at Argonne led to relating radon variations in the atmosphere to meteorological variables and more sensitive (modified Lucas technique) methods for radon determinations. The next major step in radon re-

TABLE 7.6

Highlights of the Department of Energy/OHER Radon-Related
Research Program

Year	Description of Highlight
1949	Low background ionization chambers were developed at the Environmental Measurements Laboratory and subsequently used as standard radon measurement devices.
1951	Environmental Measurements Laboratory began program to measure radon in mine atmospheres.
1952	Radon breath analysis technique for estimating radium body burdens in people was developed at Argonne National Laboratory.
1953	University of Rochester began studies of inhaled radon progeny in mice.
1956	Radon Scintillation flask counting technique developed at Los Alamos and Argonne National Laboratories.
1956	Atmospheric radon levels were related to meteorological variables at Argonne National Laboratory.
1957	Modified "Lucas" radon scintillation flask technique was developed at Argonne National Laboratory.
1965	Battelle Pacific Northwest Laboratories began inhalation studies using uranium ore dust and mine contaminants in hamsters and dogs.
1966	Environmental Measurements Laboratory began studies to characterize aerosols in uranium mine atmospheres.
1968	Argonne National Laboratory began studies of radon progeny (lead-210) in uranium miners and radium workers.
1969	Uranium miner lung cancer study based on sputum cytology began at St. Mary's Hospital, Grand Junction, CO.
1969	Environmental Measurements Laboratory began special studies of indoor radon in Tennessee, Florida, and Colorado.
1969	First comprehensive study of alpha particle dosimetry in the bronchial tree conducted at New York University.
1976	Environmental Measurements Laboratory conducted study of radon in 21 houses in the New York area.
1977	Enhanced radon exhalation after meals reported from studies by Argonne National Laboratory.
1978	Continuous radon monitor developed at New York University.

Year	Description of Highlight
1980	Lawrence Berkeley Laboratory began comprehensive multi-laboratory radon research project.
1982	Environmental Measurements Laboratory began interlaboratory radon measurement comparisons.
1984	Measurements of radon progeny in more than 200 Pennsylvania homes conducted by Argonne National Laboratory.
1984	University of New Mexico began epidemiologic study of lung cancer incidence in uranium miners.
1984	National Council on Radiation Protection and Measurements published reports on radon exposure and risk incorporating methods developed at New York University.
1985	Formal collaboration initiated between the Department of Energy/OHER and the Council of the European Communities on radon research.
1986	Argonne National Laboratory began field work on the Pennsylvania female radon lung cancer study.
1987	Department of Energy/OHER expands radon research program.

Source: Department of Energy 1987.

search came in the mid-1960s when the Battelle Pacific Northwest Laboratories began inhalation studies using uranium ore dust and mine contaminants on dogs and other research animals. Other studies continued and included investigation of uranium miners' lung cancers at Grand Junction, Colorado in 1969. Also that year, indoor radon studies were undertaken elsewhere in Colorado as well as in Tennessee and Florida. The alpha-particle dose effects on the bronchial tract were studied at New York University. In the mid- to late 1970s, several studies on indoor radon and radon in waters, soils, and other areas were undertaken. Starting around 1980, a major program was begun at Lawrence Berkeley Laboratory on a multi-laboratory radon project, and research output in several areas has continued full scale since (see table 7.6).

The Environmental Protection Agency Program. The EPA developed a strategy for indoor radon problems in 1985. In large part, this program was due to the very high radon levels found in the Reading Prong area of Pennsylvania (see introduction.) Traditionally, EPA's role in environmental concerns has been one of deciding on standards for regulation, but in this case, how can one regulate the earth? In short, EPA has undertaken to characterize risks and exposures associated with radon and to find ways to reduce radon levels in existing structures as well as guidelines to help alleviate high radon levels in future dwellings. In addition, the EPA can provide guidance, advice, and leadership to the private sector and to states in addressing the indoor radon problem.

In 1985, EPA initiated a four year program to determine levels of indoor radon across the United States. Some 50,000 homes have been involved in these studies. EPA has also taken the lead in providing useful documents for the private sector ("A Citizen's Guide to Radon," "Radon Reduction Methods: A Homemaker's Guide") as well as training programs to assist state personnel in dealing with the indoor radon problem. Mitigation techniques are also being studied, especially in Pennsylvania, by the EPA. In September 1988, the EPA attempted to stress the seriousness of the indoor radon problem. Citing recent work indicating lung cancer fatalities near 20,000 per year in the United States, the EPA now recommends all home owners have their houses tested for radon and to initiate mitigation measures in cases where the indoor level is over 4 pCi/L. The EPA is now undertaking a random study of an additional 20,000 dwellings across the United States and, in addition, is providing financial assistance to several states for radon projects.

Other Federal Programs. The Department of Housing and Urban Development (HUD) is directing its research efforts to assuring that HUD-assisted housing will have low radon levels, i.e., under 4 pCi/L. In addition, HUD is the lead agency in studying 120,000 acres of reclaimed phosphate and mined lands in Florida as well as 300,000 acres of unmined land with known phosphate deposits. Phosphates in general (see chapter 2) contain fairly high uranium and pose a long-term radon release threat. HUD is working with the state of Florida and the U.S. Steel Corporation on the project. HUD has also been active in Butte, Montana, mitigating the problem of high radon levels in public housing and constructing new public housing with acceptable radon levels.

The Bonneville Power Administration (BPA) and the Tennessee Valley Authority (TVA) are also involved with radon research. BPA offers technical assistance for home owners who have weatherized their houses using BPA guidelines by providing radon monitoring and, if necessary, mitigation. The TVA is studying homes in Alabama, Mississippi, and Tennessee to determine the effect of the use of phosphate slags in building materials versus homes where such materials were not used. If necessary, mitigation efforts will follow.

The National Cancer Institute (NCI) is sponsoring various radon research studies in the People's Republic of China, Sweden, and New Jersey. Their efforts are to determine how radon influences lung cancer.

Efforts to Detect Indoor Radon. The effort of various states to determine the indoor radon problem varies considerably. Following is a summary of some of ten different projects and approaches being undertaken at the present time. It is obvious that the efforts are greater in some regions of the United States than in others. Of note is the minimal effort being undertaken in the southwestern United States, yet, as my research suggests, indoor radon may present a great problem for this area.

The Department of Environmental Resources in Pennsylvania has been putting forth the most effort. Due to the publicity surrounding the Watras case (see introduction) and the combined efforts of EPA and the state of Pennsylvania, a large number of homes have been tested. Table 7.1, for example, shows that over 22,000 homes have been monitored for indoor radon levels, with results indicating a mean of 5.14 pCi/L. This high figure is biased because much of the testing was done in the Reading Prong area, which has high radon potential. If such areas are not considered, then the coverage for the rest of the state is modest. New Jersey has also initiated a major program on radon testing, and 4500 homes have been tested there (see table 7.1). Both Pennsylvania and New Jersey now require disclaimers before a house can be sold. In New York, over 4900 home have been tested; some counties in New York have also passed such a law.

Further south, along the East Coast, increased efforts are noted for Maryland and Virginia. Still further south, in Florida, there is a fairly active exploration of radon levels in homes built on phosphate waste lands (see chapter 2) or where phosphate-derived gypboard has been used in construction. Meanwhile, in the West, both Washington and Oregon have had active programs on radon detection. In Washington, many of the nearly 12,000 measurements were made in the Spokane

River Valley in response to concern over uraniferous rocks in that area. California has had nearly 1000 measurements.

In September 1988, the EPA announced the results of its radon sampling for several areas of the United States. In 11,000 measurements, including homes in Arizona, Indiana, Massachusetts, Minnesota, Missouri, North Dakota, and Pennsylvania, over one-third yielded radon levels above 4 pCi/L. North Dakota alone yielded 65 percent over 4 pCi/L, and Minnesota, 46 percent. These EPA results are intended to assist state governments in evaluating the potential for radon levels in their areas and to take steps to address the problem in their states.

The Environmental Improvement Division for the state of New Mexico has recently initiated a study, using charcoal detectors, of 1000 dwellings; 25% were above 4 pCi/L. It is likely that data taken from this screening exercise will have to be rechecked by long-duration measuring techniques.

Over all, members of state governments are becoming more active in tackling the indoor radon problem, but many of these efforts are Band-Aid and shotgun approaches. Some results are given below.

Maine: Studies show that winter readings are higher than summer readings; that higher indoor radon is more likely in homes built over granitic bedrock than those built over metamorphic rocks; and that some potable well waters (i.e., used for drinking) may contribute markedly to the indoor radon level.

South Carolina: Work in the Charleston area shows that, in general, most homes contain fairly low indoor radon levels, averaging about 1 pCi/L. Low outside air values are also a factor in Charleston. Only 20 homes were used in the study, however. (See readings in the Appendix.)

North Dakota: A study by Doyle and coworkers (see Nazaroff and Nero 1987) of 11 homes in the area around Fargo yielded surprisingly high indoor radon values with a geometric mean of 5.7 pCi/L. Follow-up work by the EPA in 1988 shows that, for several hundreds of homes, over 63 percent yielded indoor radon levels above 4 pCi/L. The reasons for these high levels are now being determined by detailed studies of geology, building materials, and related factors.

Colorado: Due to its high mean elevation and the abundance of granitic rocks, Colorado has the highest background radiation of any of the contiguous 48 states. Indoor radon levels have been found to be high in Colorado Springs (geometric mean of 2 pCi/L, range approximately 1 to 9 pCi/L), Boulder, and other communities. Here, the presence of granitic rocks (including granitic detritus-rich sedimentary rocks) provides a convenient source for radon. The state of Colorado, along with EPA, has initiated a study of the magnitude of the indoor radon problem.

Pennsylvania: In a study of 169 homes in the Pittsburgh area, Cohen of the University of Pittsburgh found that basements yielded average radon levels of over 6 pCi/L, first floors about 2.6 pCi/L, and second stories about 2.0 pCi/L (see Nero 1987). These data, while not surprising, do again point out the ease with which radon can enter homes with basements. Of interest also is that the second floor readings average fairly close to that for the first floor readings, indicating circulation of the radon-bearing air within the home.

COMMENTS ON THE FEDERAL RADON POLICIES

There are many inconsistencies in federal radon policies. As the EPA has pointed out, HUD will not provide insurance on loans for homes built on reclaimed phosphate lands, while it does require radon determinations on homes in the Butte, Montana, area. At the same time, HUD has no policy for the Reading Prong in Pennsylvania. Further, the Farmers Home Administration of the Department of Agriculture and the Veterans Administration have no policy on radon. The Bonneville Power Administration (BPA) will provide assistance on homes it has helped weatherize; but DOE, which has also promoted and helped weatherizing homes, has no such plan. The differences in determining what level of radon requires mitigation or attention also varies from group to group (see discussion in chapter 1). The EPA uses 4 pCi/L, above which remediation is recommended, the BPA uses 5 pCi/L, the NCRP uses 8 pCi/L, and so on. Further, permitted levels of exposure for miners over long periods of time may actually be lower than the dose received by residents in many homes.

It is very clear that there is no firm federal strategy to address the indoor radon problem. In part, this is because recognition of radon as a major health problem is new, although frankly, I am perplexed by the length of time it has taken federal agencies to come to grips with what appears as a troublesome and well-recognized problem. More important, there is the usual lack of communication among the federal agencies, along with a combination of jockeying for power and competition for funding of the expensive programs at national laboratories. At the same time, researchers from universities and the private sector are struggling for funds for their proposed work.

Other problems, discussed briefly in chapter 1, include: What are the legal ramifications of indoor radon? In the absence of federal guidelines and regulations, where does the responsibility to public health lie? Can a state, for example, mandate a radon level when the federal government has not? Could such a level be contested?

I believe that a federal commission should be established to put the various federal agencies on the same track, obtain a consensus, and establish working guidelines that will be enforced by federal regulations. Presumably, the states would also follow this plan, and the private sector would also address the radon problem. After attempts to involve 68 builders and contractors in issues concerning the radon problem potential for new homes in Albuquerque, only one builder responded and looked into the matter. Contacts with realtors produced the same poor results. Very simply, these people are in business to build and sell homes; they do not wish to deal with a potential problem.

On top of this, regulations on radon testing are also needed. A prospective buyer may be able to find what radon levels are, but can he or she be confident of the results? Federal guidelines and policy must fall into some realm of scientific accuracy and common sense.

RESPONSIBILITIES OF BUILDERS, BUYERS, AND SELLERS OF HOMES

As William Ethier, litigation counsel for the United States National Association of Home Builders, has remarked, "a home owner always has a gun called a lawsuit pointed at the builder. It is the job of builders and their attorneys to remove as many bullets from that gun as possible." (Radon News Digest, Sept. 1988: 7–9). Builders are committed to constructing homes for buyers that are safe and affordable. Yet, in the area of radon, the lack of federal and state guidelines about radon poses potentially serious obstacles to the "safe" aspect of new homes. According to Ethier, claims can arise against builders for the following reasons:

a) breach of express warranty—written contract containing statement or representation about radon, or indirect "safety" statements that could be construed to include radon;
b) breach of implied warranty—those warranties placed on builders by statutes or courts, which, since the wording is often vague, may or may not include statements that can be applied to radon;
c) negligence—which could include poor siting, poor construction, and not warning buyers of potential high radon levels;
d) fraud and misrepresentation—failure to warn about radon;
e) strict liability—the house is considered a product, and like all other faulty products, if this is the case, the maker is held liable.

These potential claim factors are sometimes difficult to prove, in a sense—but sense doesn't always turn out to be the prime factor in litigation. One could, for example, claim mental anguish, stress, etc., in some states without having to demonstrate any physical damage to receive a settlement involving radon. Further, soil testing for radon is imprecise in many instances, and a house could be sited in good faith but in an area not previously known to be high in radon. Further, what is the role of the buyer? What if the buyer smokes? Does the buyer's negligence override one of the factors that could be otherwise the builder's responsibility? What are reasonable statutes of limitations for deciding if radon is or is not a problem? These are all very real questions and, considering the many uncertainties surrounding the radon issue, ones without easy answers.

Ethier (1988) goes on to point out that builders can reduce the risks of liability in several ways. First, warranties involving radon should not be given. In the absence of legal guidelines, a warranty may prove to be invalid in any regard. Most builders include in their warranties wordage on workmanship, materials, and so on; these warranties should be specific to the point that there is no loophole for a later claim involving radon.

Second, it would be wise to include some specific statements concerning radon in clear language to thwart later potential claims against the builder for negligence, fraud, or misrepresentation. This kind of disclaimer should not be so harsh as to unnecessarily worry the prospective buyer, but it should be clear enough to protect both parties. It is advantageous to have actual radon level determinations on hand. In most cases, these can be obtained during the final months of home construction. Ethier notes that short-term radon measurements, such as those obtained by grab sampling, short-term continuous sampling, or charcoal detection, should not be used in support of the radon level; rather long-duration (1-3 months or longer) methods, such as alpha detectors, should be used. While this raises the problem of short versus delayed turnover in real estate, the protection aspect may make it worthwhile (a buyer-risk clause could arrange for possession while the testing is being done).

Third, builders should advertise and use methods designed to prevent radon entry. And fourth, builders should obtain guarantees from their subcontractors about radon level safety so that faults in this area can be, if necessary, determined for liability.

I believe that builders should take every step to minimize indoor radon while construction is in progress. A few wisely spent dollars during this period are worthwhile when one considers the expense of potential law suits. At the same time, in the absence of builder responsibility, prospective buyers should insist on radon information—including possible remedial steps—before taking possession.

Recently, the National Association of Home Builders (May 1988) has approved a greatly expanded program of research and education to determine the best techniques to use for minimizing radon entry during house construction. This is certainly the proper step to take.

Now for the thorny question of selling houses, which I have addressed briefly in the introduction and in discussions with real estate agents. Many realtors are very cool toward what they see as a potential risk to selling a house. If a house can be sold without a radon level determination, then why ask for trouble by taking measurements? If real estate agents blatantly—or out of ignorance—ignore such testing, are they, or is the previous owner or builder, etc., liable if high radon levels are found after the sale? Here, too, some legal guidelines are needed. Perhaps this might even require a grandfather clause to go into effect after some specified time, so that home owners can have measurements made and take the necessary remedial actions.

In 1987, the National Association of Realtors adopted the view that indoor testing of homes for radon levels should be encouraged, but the agency shied away from endorsing radon disclosure forms as part of sales. In Pennsylvania and New Jersey, this type of disclosure is law. In most other states no such legislation exists. The position of the NAR is that realtors are under no obligation to discover and disclose high levels of radon in houses they are marketing (i.e., out of sight, out of mind). Again, I believe that realtors should obtain such information (although just who pays for it is an unresolved question), and, if high levels are found, all parties involved should be advised. This type of honesty is so simple it is baffling that NAR does not endorse it.

For new dwellings, the solution is simple. There must be federal and state guidelines to lead the builder and buyer along a well-paved path so that radon is minimized and any excess radon levels can be determined and mitigated.

TELEPHONE NUMBERS CONSUMERS CAN CALL FOR RADON INFORMATION

The telephone numbers where consumers can call for information about radon and the names of firms/individuals that will provide radon measuring services are given in the appendix. These organizations have had their measuring devices checked in the EPA's environmental monitoring laboratory. The EPA notes that of nearly 2,000 radon service companies in the United States, only 340 have taken pains to have their services approved.

CONCLUSIONS

Extensive indoor radon measurements have been taken in the United States, with a disproportionately high number from Pennsylvania, Washington, Oregon, New York, New Jersey, and Maine (over 50,000 of the 61,000 reported by Alter and Oswald 1987). Researchers have reported a very high number of dwellings, above 20 percent, with readings over 4 pCi/L, while some writers argue for a figure closer to 7 percent. My own limited work in New Mexico, where the soil is immature due to the arid conditions, agrees with the higher figure (i.e., about 28 percent). Since the measurements have not been very systematic, it is not clear how many dwellings in the United States are in fact over 4 pCi/L. The EPA, with the DOE and NRC, must come to grips with this problem and implement proper sampling procedures which would follow proper pre-sampling studies (usually not done). While it is beneficial to the national laboratories to have large amounts of money available for sophisticated radon studies, it is still not certain that the basic question of how best to provide a statistically meaningful sample for the United States can be obtained. Geologic favorability maps suggest the possible presence of uranium over much of the country, including the southern states, and it is my experience that most of the arid lands, which include some large to very large metropolitan areas, have been ignored in the programs to date.

8

Remedial Action Methods in the United States

Through the complex interplay of radon sources such as soil, water, building materials, and air and factors affecting these—permeability, porosity, and moisture content of soils, humidity, wind velocity, pressure, temperature, and others—some dwellings yield indoor radon values of 4 pCi/L or above, but most do not. The mean value for some 1250 dwellings in the United States is 1 pCi/L, well under the EPA recommended value of 4 pCI/L or above. I have noted that for the Albuquerque area, where many homes are built on granitic detritus from the nearby Sandia Mountains and where soil uranium and radium are essentially the same as in the parent granite, some 200 dwellings yield an arithmetic average of 3.3 pCi/L and an arithmetic mean of 2.6 pCi/L. My results are even more significant in that the detectors were deployed in living areas as opposed to isolated areas.

The basic question is how to lower levels to below 4 pCi/L. Even at this level there is still risk, however (see chapter 1). Some of the methods proposed by EPA and others are described below.

VENTILATION TO THE OUTSIDE

When people voice concern about indoor radon, a rule of thumb that more often than not works is to ask. "Do you feel a draft anywhere in the house?" If so, the chances are that the indoor air effectively and efficiently exchanges with the outside air at a sufficiently rapid pace to flush out the radon from the house. One can induce the same effect by opening a window or using a simple window fan, as illustrated in figures 8.1 and 8.2. In figure 8.2, air from outside is forced into the house and exhausted through either open windows, down draft from the inlet window, or a fan. This is intentional, for one would not want to reverse these conditions because this would increase the pressure gradient that already exists, and even more radon would be pulled into the house from the soil, building materials, etc. Since most homes have an air exchange rate of 1.0 ach (air changes per hour), opening windows tends to increase this exchange. For very tight homes, the rate of exchange is sometimes only 0.1 ach, thus increasing the possible accumulation of indoor radon.

There is somewhat of a paradox here. Many home owners, often with the encouragement and assistance of utility companies making a positive

FIGURE 8.1
Radon reduction by passive means. Opening windows as indicated helps flush out indoor radon. From EPA (1986).

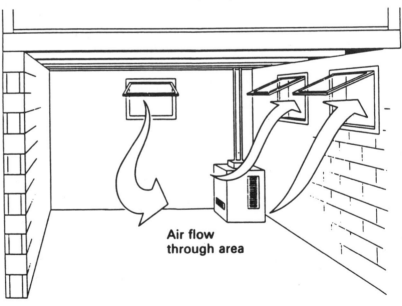

Air flow
through area

FIGURE 8.2

Radon reduction by use of window fan. Forced air entry via fan helps flush out indoor radon. From EPA (1986).

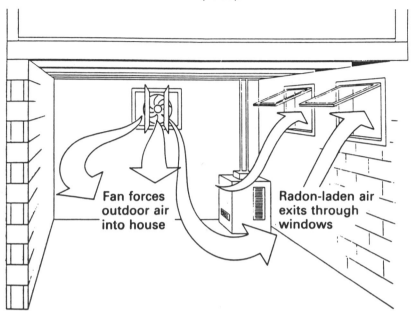

Fan forces outdoor air into house

Radon-laden air exits through windows

public relations gesture, have taken pains to lower heating and cooling bills by overinsulating their homes. The result of this overinsulation is that the homes then become traps for radon and other gases. In the cases of homes with active solar capacity, where the air is recirculated instead of exchanged with the outside air, this effect is magnified; thus, potential radon (and other) gas accumulation is further increased .

For tightly constructed homes with a somewhat high indoor radon level, the use of fan-induced or natural ventilation is often a straightforward method for lowering indoor radon levels. The negative side, in some opinions, is that heating costs in the winter and cooling costs in the summer increase. In my opinion, this is more than offset: Lowering radon levels to a more healthy level is more important than minor financial considerations. In the cases where the overall house indoor level is more than 30 pCi/L, even forced natural ventilation may not be sufficient to lower the indoor radon levels. In these cases, more stringent steps must be taken.

There are many so-called common sense aspects of natural and forced ventilation, however. Wind direction and strength may change rapidly, thus negating the forced ventilation effect. Precipitation may also be a factor. Simply being aware of these conditions and taking pains to ensure that the flow of radon promoted is away from the indoors are sufficient in most cases.

Basements or crawl spaces are obviously the areas that should be ventilated, and it is further recommended that such areas not be used for family activities. If the house is of a simple slab design, then the living areas must be ventilated. Again, as far as indoor radon goes, outside air is good for you. Keeping windows open whenever possible is highly recommended. If, however, by ventilating a crawl space or basement, means that pipes or other weather-sensitive items will be exposed to severe cold, then they must be insulated to prevent freezing.

VENTILATION BY HEAT-RECOVERY METHODS

This method uses a heat recovery ventilator as shown in figure 8.3. Here the air from the room, a basement as shown, is used to heat incoming air in the winter and to cool incoming air in the summer. The heated or cooled air then circulates as indicated.

Recovery methods are more sophisticated than the natural or forced ventilation methods described in the preceding section (see figs. 8.1 and 8.2). The costs are variable but may be as high as $2000, and heating and cooling bills in winter and summer, respectively, will increase. These ventilation units are run continuously for best results. For houses without basements or crawl spaces, the units must be installed in living areas. If the entry points of radon have been determined, then a simple window unit near such points may suffice.

AIR SUPPLY

Many appliances such as dishwashers, washing machines, dryers, and even woodstoves and fireplaces draw air from the inside of the house and vent it to the outside. As discussed earlier in this book, this is an unfavorable pressure gradient situation, something like putting a fan in the window to draw the inside air out, and it creates a positive pressure for radon to more readily emanate into the structure. Figure 8.4 shows

FIGURE 8.3

Radon reduction by heat recovery ventilator. See text for detail. From EPA (1986).

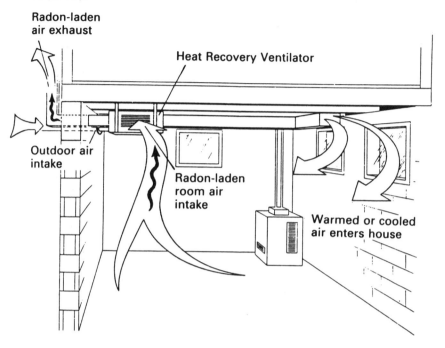

some of these items and indicates that extra vents from the outside are placed so that air is drawn into the units from the outside, thus preventing an unfavorable pressure gradient for radon inside the house.

The cost of such renovations is difficult to predict. Usually, however, the types of exhaust systems needed can be installed rather inexpensively. Other expenses come from running washing machines and dryers because more energy is needed to heat the air. This, however, is a justified expense if it lowers the radon levels. How much the levels can be lowered is difficult to assess, since homes are so variable, but very good results have been reported (EPA 1986).

EXPOSED EARTH COVERINGS

In many homes, basements have earthen floors; in others, special features such as sunken bathtubs, pump housings, furnace foundations,

FIGURE 8.4
Radon reduction by extra venting to compensate for radon release due to furnaces, washers and dryers, etc. See text for detail. From EPA (1986).

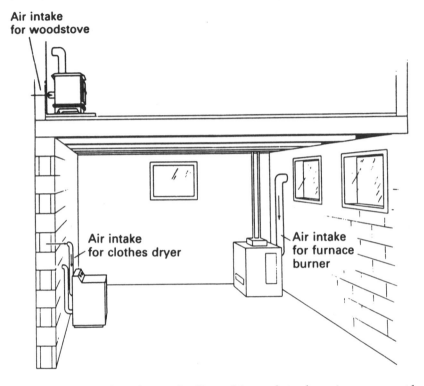

etc., are anchored in the earth. Since this earth is the prime source of radon for most homes, it is necessary to cover these exposures by some method.

The EPA recommends covering earthen floors with cement or, in some cases, gas-proof plastic liners to prevent radon entry. These coverings are usually sufficient to prevent much of the readily available radon from entering. The venting system for a hypothetical sump-in-basement is shown in figure 8.5. Installing the covering takes skill, and professional contractors should be used if there is any question about the homeowner's expertise in such matters.

The cost of operating the venting systems, including small exhaust fans to promote radon removal, is low. The fee for the covering, includ-

FIGURE 8.5
Radon reduction by sump-in-basement. See text for detail. From EPA (1986).

Outside fan
draws radon
away from house

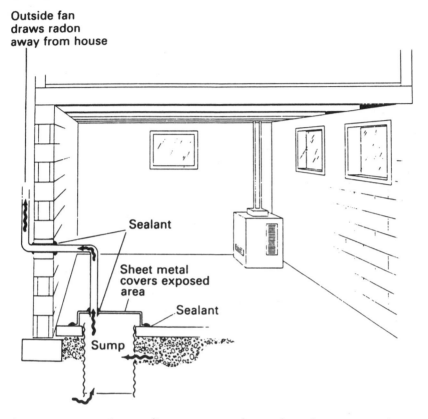

Sealant

Sheet metal
covers exposed
area

Sealant

Sump

ing extra excavation, sealings, etc., may be moderately expensive, however. In addition, care must be taken to periodically inspect the new covering for cracks or other effects from settling foundations.

CRACKS AND OPENINGS

In many homes, the entry points for radon are flaws in the original construction or in some postprimary construction phase. Figure 8.6 shows a view of a typical basement (or first floor room) with many areas of entry including cracks in walls, cracks in the floor, separation of floor and ceiling from walls, cracks around pipes, and so on. All such cracks

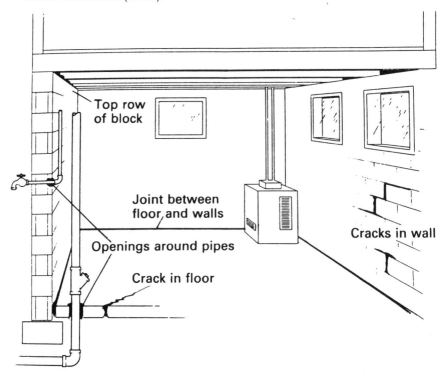

are potential inlets for radon, especially in basements. While some cracks
and other openings can be visible, other cracks and/or inlets due to
weakened material are more difficult to assess. Experienced contractors,
using modern equipment, should be able to determine these more subtle
openings, but the cost for locating and subsequent sealing may be fairly
high.

With time, houses settle, mortar and other sealants degrade, and more
openings may result. Thus, continued monitoring of the house is necessary.

The EPA is careful to point out that in cases of very high indoor
radon levels, even sealing these cracks may not be enough. If there is an
extremely high radon source in the soil or rock under or around the
structure, then a high radon flux into the house may result. In such a

case, ways to remove the radon from the surrounding soil may be necessary (discussed below).

DRAIN-TILE SUCTION

In many homes, water is drained away from the structure by use of perforated pipes called drain tiles. If such drain tiles surround a house, they can be used to draw radon away from the house. A typical drain tile system is shown in figure 8.7.

These systems work well in areas of moderate to high total precipitation and can be as much as 98 percent efficient in removing soil radon away from a home. In the method shown in figure 8.7, an exhaust fan is linked to the drain tile system to remove the radon.

This system, although expensive, can be very efficient in areas where the soil moisture is high. Expenditure of perhaps $2000, plus some energy operating costs, should be anticipated.

VENTILATION OF BLOCK WALLS

As described in chapter 5, block walls are often major conduits for indoor radon. Even floor covering, drain tiles, etc., can have only minimally successful effects in lowering indoor radon if the radon is entering from the block walls.

The approaches for treating this problem are shown in figures 8.8 and 8.9. The pipe-in-wall method (see fig. 8.8) relies on simply placing a pipe into the block wall from which radon is pumped via a fan to the outside. With proper seals, this method can easily remove much of the indoor radon from the block walls.

For the baseboard approach (fig. 8.9), sheet metal is installed around the floor, and holes are drilled into the block walls. The radon is then pulled from the block walls and vented to the outside as indicated.

Costs for installing the pipes in the walls may run $2000 to $3000, and the baseboard method may cost about $5000 (EPA 1986), yet these methods, expensive as they seem, may be necessary if block walls are the prime source of indoor radon.

FIGURE 8.7
Radon reduction by drain tile suction. See text for detail. From EPA (1986).

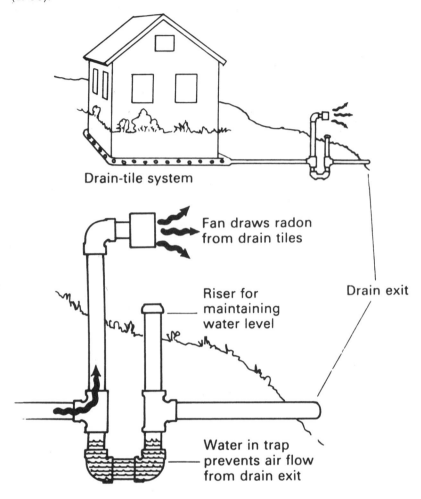

Drain-tile system

Fan draws radon from drain tiles

Riser for maintaining water level

Drain exit

Water in trap prevents air flow from drain exit

SUB-SLAB SUCTION

The sub-slab suction method is illustrated in figure 8.10. Here pipes are drilled through the floor of a house sited directly on a slab of cement or concrete over soil. The pipes are vented via an exhaust fan as indicated

FIGURE 8.8

Radon reduction by pipe-in-wall method. See text for detail. From EPA (1986).

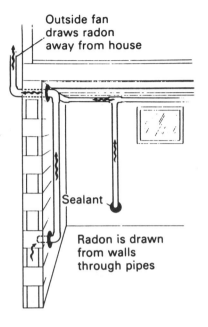

Outside fan
draws radon
away from house

Sealant

Radon is drawn
from walls
through pipes

Pipe-In-Wall Approach

in figure 8.10. Any radon that accumulates in the aggregate over the soil and below the floor is then sucked into the pipes and exhausted away from the home. The EPA is careful to note that if the homes contain hollow block walls, the sub-slab venting method may not be sufficient to lower radon levels. In such cases, use of pipe-in-wall or baseboard hollow block wall methods must be employed.

The expense for the sub-slab venting, sealing, and exhaust systems may run as high as $3000, but homeowners can do it themselves for about $300.

Sub-slab ventilation methods have proven to be remarkably successful. Dr. Douglas Mose of George Mason University has reported over 95 percent success in lowering radon levels to under 3 pCi/L in parts of

FIGURE 8.9

Radon reduction by baseboard approach. See text for detail. From EPA (1986).

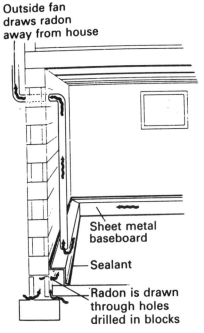

Outside fan
draws radon
away from house

Sheet metal
baseboard

Sealant

Radon is drawn
through holes
drilled in blocks

Baseboard Approach

Virginia and Maryland. Since the values before mitigation showed 80 percent of the homes over 3 pCi/L, the results are striking. A simple fan-exhaust system is used to vent the sub-slab radon to the atmosphere.

RADON DAUGHTER REMOVAL FROM AIR

It is also possible to remove much of the radon daughter material from the air. This can be done by several devices, including filtering, increased plateout (i.e., deposit of the radon daughters onto surfaces) by air circulation, and electric methods. Use of filters is not always successful, yet it has been reported that for many dwellings, the use of filters reduced the average dose by amounts ranging from 20 to 50 percent for 0.3 to 2 air exchanges through the filters per hour. The success of filters depends on

FIGURE 8.10
Radon reduction by sub-slab suction. See text for detail. From EPA (1986).

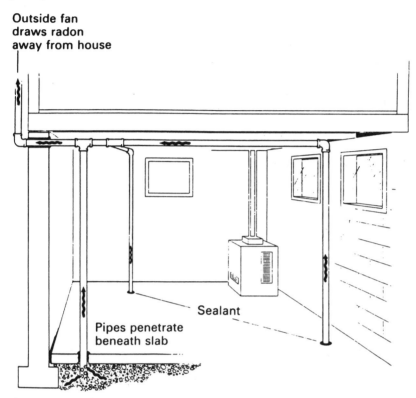

Outside fan draws radon away from house

Sealant

Pipes penetrate beneath slab

many factors, including the volume and surface area of the room, the aerosol characteristics of the room (i.e., dusty versus nondusty, etc.), and the rate of filtration. For higher filtration rates, such as 2 to 3 exchanges per hour, the removal of potential alpha-energy concentration is about 90 percent.

Radon daughters are often referred to as either "attached" or "unattached." In the former case, the radon daughters, and especially charged ions such as those of [218]Po, are fixed to some other material. In the unattached case, the radon daughters remain free. When a fan was used, the plateout rate of unattached particles increased from eight per hour to 30 to 50 per hour. Researchers point out that there is a large variation in the results of such studies, and it would be unwise to rely solely on fans to lower the concentration of radon daughters in any living space.

In electric, or, more properly, electric field, methods, the advantage is taken of the fact that many radon daughters, appearing as attached ions, are charged. Since ^{218}Po is positively charged, it makes sense to try to remove this isotope by an electrically negative plate or wire. Since the distance from the charged wire or plate to the charged radon daughters is critical, large rooms may have to have many of these devices. For example, lead (Pb) and bismuth (Bi) isotopes must be within 50 to 60 cm of the wire or plate, but polonium (Po) isotopes must be within 20 cm of the wire. Hence, these methods are somewhat cumbersome in application. It has been noted further that use of an electrofilter in conjunction with the wire or plate results in about double the amount of radon daughters removed, but even with this improvement, only 50 to 80 percent removal within the volume close to the wire or plate is possible. Studies to integrate fans to push more air toward the collectors and filters have not been carried out.

In short, the methods to remove radon daughters from the air in dwellings are somewhat tenuous and should be employed only if all other methods to prevent entry of radon gas have failed.

CONCLUSIONS

The remedial actions necessary to mitigate the indoor radon problem have been described briefly in this chapter. Some of the methods are more sophisticated than others. Obviously, before any remedial plan can be undertaken, the question of whether there is a problem of high indoor radon level in a given house must be answered.

At the beginning of the chapter, it was noted that the common sense statement about "feeling a draft" is still a good rule of thumb for indicating likelihood of high radon levels. That is, in drafty houses, radon levels can be low. However, in newer, energy-efficient houses built with extra insulation, active solar systems, and the like, indoor radon monitoring is essential.

I have taken great pains to document the existence of indoor radon levels in several homes in Delaware, Pennsylvania, and New Mexico, and the owners were advised of ways in which to reduce the radon levels. In several cases, passive and forced ventilation worked. In another case, running only the fan part of a heating-cooling system for one hour a day was sufficient. In two other situations, forced venting by use of a pipe-pump system worked.

The bottom line is one of personal choice. Since there are no federal or state guidelines, and since the actual remediation may be time consuming and expensive, homeowners must make up their minds whether the levels determined warrant remedial attention. My opinion is that it would be foolhardy not to mitigate the radon.

9

Radon Studies in Countries Outside the United States

While the main focus on indoor radon problems in this book is for the United States, the dilemma, of course, is of worldwide interest and concern. Detailed studies have been carried out in a number of countries throughout the world, with special interest in Europe and Canada. For convenience, I break these down by individual country, then summarize the overall findings.

In the United States, there is a pronounced correlation between indoor radon and radon daughter concentrations and uranium content, permeability, and moisture content of local rocks. These factors exist worldwide and will be commented on in turn.

EUROPE

Sweden. Much of the pioneering work in tracing indoor radon has been carried out in the Scandinavian countries and especially in Sweden. Akerblom (1986) notes that in Sweden, perhaps 400 to 800 deaths (in a population of eight million) from lung cancer that occur each year can be attributed to indoor radon. At first glance, Sweden's problem is in part due to the presence of uraniferous granites and very high concentrations of uranium-bearing Alum Shale. Mention has been made in the introduction of the influence of Alum Shale and its use in concrete manufacturing as a contributing source for indoor radon. Swedish investigators point out that not only are areas of high uranium in rocks potentially high radon risk areas, but other rocks and rock materials are also of concern. Thus, in weathered, fractured, uranium-rich granites, the radon release is high; yet in some Alum Shale, despite high uranium, the low permeability of the shale retards radon release. It is in the high pore space glacial eskers where radon release is an extreme threat.

Other workers have summarized many of the earlier radon studies in Sweden. They report that for 315 detached houses built before 1975, an average indoor radon value of 3.3 pCi/L was obtained, while the value was 2.3 pCi/L for 191 apartments built before 1975. Of these, 2 percent of the houses had radon values of 20 pCi/L, as did 0.5 percent of the apartments. The level for remedial action in Sweden is 10.7 pCi/L equilibrium-equivalent daughter concentration (EEDC), assuming 0.5 as the equilibrium factor between radon and its daughters. These workers also report on post-1975 structures with approximately the same results, although a smaller number of houses fall in the above-20 pCi/L range. Recently, some 60,000 dwellings have been monitored, and the investigators note that 2 percent of all these still fall above 10 pCi/L for radon daughters. Most of the high radon is due to soil and rock radon, although in some cases, building materials may play an important role. The importance of soil/rock radon is summarized in table 9.1. Here it is noted that a very wide range of values has been observed. For sands and

TABLE 9.1
Soil and Rock Radon Levels in Sweden

Soil or Rock	$^{226}Ra(pCi/kg)$[a]	$^{222}Rn(pCi/L)$[b]
Till, normal	405–1,675	135–811
Till, with granite	810–3,378	270–1,622
Till, with U-granite	3,378–9,730	270–5,405
Esker, gravel	811–2,027	270–4,504
Sand, silt	162–2,027	54–811
Clay	676–2,702	270–2,162
Sils with Alum Shale	4,730–67,567	1,350–27,000

[a] 332 pCi/kg ^{226}Ra is equivalent to 1 ppm uraniun.
[b] Soil air measured at one meter depth in all cases.
Source: Akerbolm 1986.

silts removed from uraniferous areas, the range is from 54 pCi/L to 811 pCi/L, while for soils with Alum Shale, the range is a staggering 1350 pCi/L to over 27,000 pCi/L. For some perspective, soils in the United States average about 100 pCi/L at a depth of 38 cm. The Swedish government is fully aware that indoor radon is a significant health problem and is taking steps to remedy it.

Finland. Finland has not only uraniferous rocks with the same high-permeability glacial materials as Sweden but also other similarities to Sweden. Investigators there report that for over 2000 dwellings in Finland, the mean is 1.73 pCi/L, with about 2 percent of the dwellings over 20 pCi/L. The Finnish results are, as expected, very similar to those in Sweden. While Finland does not contain shales as rich in uranium as the Alum Shale in Sweden, it does contain uraniferous granites, pegmatites, and glacial deposits. As has been pointed out above, since Alum Shale is rich in uranium and fairly impermeable, much of the radon is not readily released from it.

Denmark and Norway. Studies in Denmark and Norway are not as complete as those in Sweden and Finland. It has been noted that summer readings in Denmark average only 0.8 pCi/L and winter readings average 2.4 pCi/L, but these are pilot study results only. Norway's survey is currently underway. Due to the presence of uranium-rich granitic rocks as in Sweden, high background soil and rock radon is to be expected.

Federal Republic of Germany. German scientists have reported on indoor radon measurements in over 6000 homes for the Federal Republic of Germany and found a gross mean of 1.1 pCi/L, with only 0.2 percent of the homes noted above 13.5 pCi/L. Open air radon values in the Federal Republic of Germany range from 0.3 to 0.45 pCi/L, with a mean of 0.16 pCi/L, according to other German scientists. They point out that for total dose of radon daughters to individuals, the outdoor air adds a small but not insignificant radon daughter contribution to the lungs. Further, these workers point out that many German homes are built with well-sealed basements to keep out moisture; hence, radon contributions from soils are lower than they might be otherwise. Here the indoor mean of 1.1 pCi/L is lower than in countries such as Sweden that have a greater soil-rock contribution. Further, mine and mill tailings are not commonly used as building materials for dwellings. Finally, these scientists also show that the equilibrium factor for radon daughters is 0.3 and 0.4 for inside air.

The Netherlands. Located on the edge of the North Sea Basin, the Netherlands presents a special situation. Unlike the United States, Canada, and Sweden, the Netherlands is largely situated on uncemented sediments of young geologic age that persist to several hundred meters. These materials are not very uraniferous, and soil radon emanation rates are low. Because bricks are made from these local, uranium-poor materials, building materials are also low in radon emanation. Further, Dutch scientists point out that the climate is moderate and somewhat windy so that outdoor air is very low in radon. High rates of ventilation inside homes help reduce radon levels. Other Dutch scientists report that 1000 dwellings in the Netherlands yield a grand mean of .65 pCi/L with a lognormal distribution, but fewer high levels (i.e., above 10 pCi/L) than in other European countries. As expected, homes with double pane windows yield slightly higher radon values than those with single panes, homes with concrete floors yield higher values than those with wooden floors, and the lower levels of houses yield higher values than upper levels.

The United Kingdom. British scientists have reported on indoor radon measurements in over 2000 homes in England. The grand mean for these measurements is only 0.41 pCi/L, although areas such as Lands End at Cornwall yield higher values. This is because many of the granitic rocks in the Cornwall area are uraniferous (it should also be noted that when plate tectonics are considered, the granitic rocks of the Cornwall area and the uraniferous granitic rocks of the Maritime Provinces and the northeastern United States were all once part of the same land province). Most basements in the England are well sealed to keep out moisture, and this further reduces the indoor radon values. Summer readings are about half those of the winter readings. The Cornwall area yields an average indoor radon value of 10.5 pCi/L, over fifteen times the average for the rest of the England. Studies from Scotland and Ireland are in progress at the present time; already investigators report that 10 percent of 250 homes there yield indoor radon values over 3 pCi/L and a geometric mean of 1.16 pCi/L.

Other European Countries. Studies of indoor radon from other European countries are either in progress, not being carried out, or being planned. Some local studies are worth noting, however. In Yugoslavia, attention has been paid to the Slovenia region, where Karst topography (dissolution features in limestone) has resulted in numerous caves, many of which are open for tourists. Scientists from Yugoslavia report radon values ranging from 2 to 150 pCi/L in these caves, with many readings in excess of 10 pCi/L. They note that very high radon levels are a real health hazard for workers in these caves. Data are not given for dwellings in the Slovenia region, however.

As part of a preliminary pilot study in Belgium, 79 dwellings yield a geometric mean of 1.1 pCi/L. In France 765 homes yield a geometric mean of 1.2 pCi/L, with 2 percent of those over 5.4 pCi/L. More data are being gathered in both countries.

CANADA

Nearly 10,000 homes from 14 major Canadian cities were sampled during June-August in the late 1970s by grab-sampling techniques. A range in values from 0.14 to 0.97 pCi/L was obtained, but this range is probably too low since the sampling period was in the summer. The investigators conducted a follow-up to this study and expanded the number of dwellings to 14,000 and the number of cities to 18. Again, the surveys were conducted for only the June-August period. Their data are summarized in table 9.2. There is a distinct geographic variation, with Winnipeg yielding the highest geometric mean of 1.54 pCi/L.

Detailed studies of cities where radon or radioactivity are suspected to be contributing factors to ill health were excluded from the Canadian study of Letourneau and his coworkers (1985). The 18 cities covered and the 14,000 dwellings sampled represent some 10,000,000 Canadians, however.

In Port Hope, North West Territory, Canada, both radium and uranium refineries have been in operation since the 1930s. The entire 2960 dwellings in Port Hope were surveyed and 550 were found to be above 2 pCi/L. Remedial steps were taken to locate and remove the radioactive materials from these homes. Contaminated building materials and wastes had been used for construction and other purposes all over the city.

In addition, uranium mining communities in Canada including Ban-

TABLE 9.2

Indoor Radon Survey Results: Canada

Location	No. Homes	Radon Geometric Mean (pCi/L)
Winnipeg, Man.	563	1.54
Regina, Sask.	961	1.33
Sudbury, Ont.	772	0.58
Brandon, Man.	561	0.84
Saskatoon, Sask.	770	0.42
Fredericton, NB	455	0.66
Edmonton, Alb.	603	0.46
Thunder Bay, Ont.	627	0.54
Sherbrooke, Que.	905	0.36
Calgary, Alb.	900	0.31
Charlottetown, PEI	814	0.41
Toronto, Ont.	751	0.31
Saint John, NB	867	0.27
St. Lawrence, Nfdl.	435	0.88
St. John's, Nfdl.	585	0.30
Montreal, Que.	600	0.29
Quebec, Que.	584	0.28
Vancouver, BC	823	0.14

Source: Letourneau et al. 1985.

croft, Elliot Lake, and Uranium City, have been extensively surveyed. The number of homes sampled were: Bancroft, 1162, Elliot Lake, 1921, and Uranium City, 632. Radon daughter levels were found to be higher in the homes in these cities than in the 18 major cities, and the data provided instigated cleanup actions.

The survey conducted in Saskatchewan is of special interest. This province, which is underlain in the north by the Precambrian Shield and in the south by the highly radioactive Ravenscrag Formation, a coaly unit, is host to many major uranium deposits. Studies covered 613 homes from 35 communities. Significant radon and radon daughter levels were noted, and this information has been used for remedial action and for guidelines for new dwellings.

JAPAN

Results from Japan are fragmentary. A geometric mean of 0.5 pCi/L was reported for 251 homes from four cities, with 2 percent above 3 pCi/L.

CONCLUSIONS

In conclusion, I note that radon studies are being carried out in many developed countries but in very few third world nations; however, in the United States, Sweden, and Canada the effort to detect and control radon levels is greater than in other developed nations. Very simply, the importance and magnitude of the indoor radon problem has not been recognized in much of the world community. This fact may have adverse health effects in the future.

10

Economic Uses of Radon Emanations

So far in this book, the emphasis has been either directly or indirectly on the indoor radon problem. Now, however, I shall examine some of the uses of radon emanations for mineral exploration and for other purposes.

For many years, geologists and other scientists have noted that there exists a high correlation of radon emanations with different geologic phenomena, including uranium deposits, geologic structures, geothermal areas, and others. In the following pages I will discuss these and some other uses of radon emanations.

URANIUM DEPOSIT PROSPECTING

The parent isotope of ^{222}Rn is ^{238}U. Hence, in areas of higher than background uranium, potential for higher than background radon flux is present. When uranium is enriched to the point of forming economic deposits, then radon, too, is present in amounts many times over background. In uranium mining, for example, mines must be constantly ventilated to flush out radon and radon daughters, and miners usually wear respirator masks and are monitored for their radiation and radon exposure. In chapter 4, it was pointed out that adverse health effects of

FIGURE 10.1a

Radon release from hypothetical uranium ore body. See text for detail.

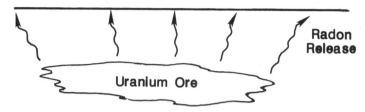

FIGURE 10.1b

Radon release from hypothetical uranium ore body, plan view. See text for detail.

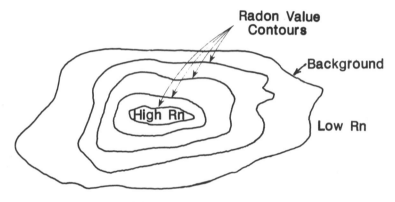

high radon are noted for many different kinds of miners: tin, gold, fluorspar, and iron as well as uranium.

Uranium prospecting can be carried out by the use of radon testing. Figure 10.1 shows a hypothetical series of uranium occurrences. In figure 10.1a, the rock and soil overlying the buried uranium deposit is favorable for radon loss, and, in plan view (see fig. 10.1b), the radon will form a series of haloes around the surface expression of the subsurface deposit. This makes drilling for buried uranium ore somewhat easier than without this kind of target. In figure 10.2, another hypothetical deposit where overlying rocks tend to prevent direct transfer of radon from deposit to the surface is shown. In such cases, the surface signature may be diffuse or not defined at all. Geologists working the area will, however, note that there is an above-average radon flux in the outcrop

FIGURE 10.2
Radon release from uranium ore in tilted strata. See text for detail.

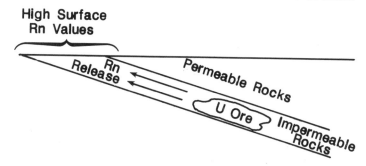

area for the permeable unit shown in figure 10.2, which will at least give a direction in which to pursue further exploration.

In young sandstone types of deposits, uranium is commonly found in zones where chemically oxidizing conditions change to reducing conditions so that U^{6+} in solution (see chapter 2) is reduced to insoluble U^{4+}. Slight loss of intermediate radioactive uranium daughters, such as ^{230}Th, cause a build up of ^{226}Ra, which is fixed in the rocks. This is the source of the ^{222}Rn. In simple occurrences, such as shown in figure 10.3, a radon halo may occur over the ore. Caution must be used here as if continued groundwater flow destroys part of the ore, the radium is still left where first found, although new radium will follow the movement of the redox front (see chapter 2). A diffuse, or smeared-out, signal may then result (see fig. 10.4).

When uranium occurs in deep deposits, it may be difficult or impos-

FIGURE 10.3
Radon release from uranium roll deposit. See text for detail.

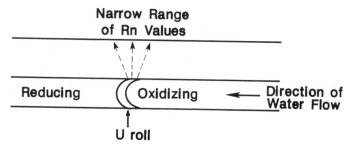

ECONOMIC USES OF RADON EMANATIONS   175

FIGURE 10.4

Radon release from remobilized uranium roll deposit. See text for detail.

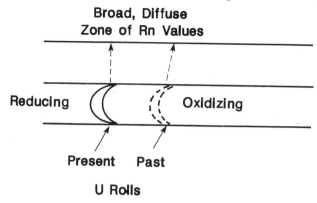

sible to detect radon at the surface unless the background radon values are rigorously known. It has also been pointed out that for the Baker Lake uranium deposit in the Northern Territories, Canada, alpha detectors buried in the early fall, left in the frozen ground over the winter, and recovered in the late spring, nicely contour on the surface over the subsurface ore deposits (see Gingrich 1983; fig. 8). These detectors were deployed in a systematic grid, and background radon was reliably established. Because the detectors were deployed for over six months, the data were very precise, and subtle differences were recognized. There are numerous other examples of radon surface data contouring over subsurface uranium deposits, and the method has been widely used for reliable uranium exploration. Figure 10.5 shows a redrawn sketch of part of the Baker Lake radon occurrence map.

False anomalies may occur, however. Figure 10.6 depicts the formation of a convection cell caused by a temperature gradient involving circulating fluids over an impermeable rock. The result is more pronounced in the winter than in the summer because a stagnant layer will form just below the surface in the summer, but not in the winter. The result, then, for the winter months is alternating high and low radon emanations. In the absence of other criteria, this could be mistakenly interpreted as uraniferous rock for the high radon signals and uranium-depleted rock for the low radon signals. Figure 10.7 illustrates how this convection cell system is blocked by the stagnant summer layer.

FIGURE 10.5
Example of radon monitoring—contouring survey. See text for detail.

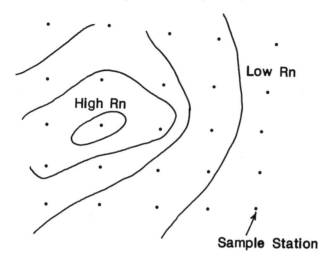

The question, then, for uranium exploration is simply when to follow or when not to follow positive radon anomalies. Here geologic expertise plays a large part. In areas of known or suspected uranium potential, positive radon anomalies usually result from carefully designed grid-type radon detector deployment; hence, any anomaly is of potential interest. Positive radon anomalies in areas known or considered to be of low uranium potential are thus likely to be due to convection cells, faults, or other factors. These, too, may be of interest for a wide variety of reasons as discussed below.

FIGURE 10.6
Effect of fluid convection on radon release for winter conditions. See text for detail.

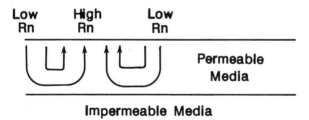

FIGURE 10.7
Effect of fluid convection on radon release for summer conditions. See text for detail.

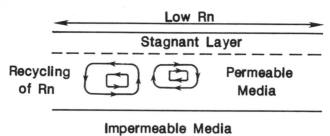

PROSPECTING FOR GEOTHERMAL RESOURCE AREAS

Geothermal energy is a minor contributor to United States and world energy, yet it is becoming more and more important. A known geothermal resource area (KGRA) usually covers a wide area. This is often because the KGRA overlies a shallow magma body, is on the flanks of a cooling volcanic pile, or is located over a large and complex near-surface hydrothermal hot-spring system. Siting of drill holes for testing for steam potential is expensive, and ways to carefully focus on small subareas of the KGRA are sought. Exploratory drill holes may cost as much as $500,000 each, so ways to reduce the number of such holes are of extreme economic importance.

Radon is commonly enriched in hydrothermal systems, and when steam is vented to the atmosphere, very high radon contents are observed. For nonvented steam, radon may act as a tracer to indicate where larger concentrations of steam are located. In the Waireki geothermal field in New Zealand, a radon study of the Crater of the Moon subarea showed marked radon positive anomalies along several points of a complex fault system. This, in turn, allowed direction of the drilling program to focus on these areas.

In brief, KGRAs are usually located initially by observation of hot springs, evidence for above-average heat flow, or a combination of these and other factors. The use of radon emanations to assist the explorationist in zeroing in on favorable targets is the prime application of the method to the discovery of new geothermal energy.

DETECTING FAULTS

A more recent use of radon emanations is for detecting faults. This can be part of earthquake prediction/protection programs and siting programs for power plants and waste disposal areas. In many areas, soil and vegetative cover mask traces of faults. It is important for siting that where foundation material stability is a requisite criterion for site safety, as accurate as possible an understanding of the subsurface geology be known. Faults are obviously important in this regard.

Like many other gases, radon gas diffuses along the planes of faults (see fig. 10.8). Unlike other gases, however, radon, because it is radioactive, can be detected by alpha-track detectors or some other means (see chapter 6). Where radon detectors have been deployed in areas of known faults, an excellent degree of correlation with radon-positive anomalies and fault location is noted. This means, for example, that most if not all sites can be studied for their radon emanation to test for the presence of faults prior to any actual construction. This is an added safety measure and should be very important. In the hypothetical cases of chemical waste disposal sites or domestic refuse sites, previously undetected faults could be noted by radon indicators; this would then raise concerns about how such faults interact with subsurface water flow. Additional study

FIGURE 10.8
Effect of faults on radon release. See text for detail.

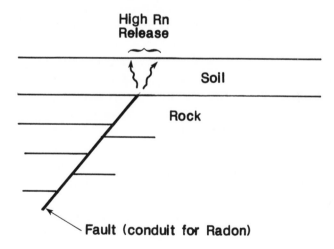

would demonstrate that the faults would not adversely effect the site; at worst, the site would be found to be unsuitable for the intended purpose. Again, it is economically and environmentally advantageous to know such information before facilities are built.

In the area of earthquake prediction, it is suspected that radon and other gases may move more readily just prior to major faulting. It is during the pre-earthquake period that subtle vibrations cause animals to become visibly affected. Above-background radon gas emissions, if measured in a continuous mode, provide the same information in many cases. Unfortunately, the number of monitoring stations required to use radon emission is ridiculously large. There is, of course, the added problem of determining whether there is a direct correlation of intensity of the positive radon anomaly with the fault intensity. This and other key questions still await the research necessary to provide the answers.

The U.S. Geological Survey is, however, conducting a study of this important topic. It reports some success in correlating radon emissions along parts of the San Andreas fault with magnitude 4.0 or greater (Richter scale) between Santa Rosa and Cholame, California. It is hoped that these studies will be expanded and accelerated in the near future.

PROSPECTING FOR PETROLEUM AND NATURAL GAS

It has been known for a very long time that certain gases, such as helium and often nitrogen, are enriched in brines associated with petroleum and natural gas. While the reasons for this association are beyond the scope of this book (and there are certainly many hypotheses from which to choose), the fact that the association exists is of importance. Radon is also enriched in these brines. As an inert gas under normal terrestrial conditions, radon does not interact with other substances, and, if lost from the brines, it will attempt to diffuse to the surface.

It has been noted that radon surveys over known oil and gas fields as well as over some suspected fields yield positive anomalies over the locations (in the subsurface) of the fields, even when the depths to the fields are on the orders of 1 to 2 kilometers. Further, investigators note that there may be a correlation of decreasing intensity of the radon anomaly with the depletion of the field as production continues.

The explorationist can make use of both facts. In defining areas for drilling, areas of positive radon anomalies may be better risks than those without such anomalies. These decisions, of course, are made with complete and on-going geologic, geophysical, and hydrologic studies. As the rate of production increases, it is always important to be prepared for a field's becoming no longer economical to drill. Radon now offers a possible way to predict this, and plans for the field can be made with adequate leadtime (i.e., will secondary recovery be attempted, will other methods be employed, or will the field be abandoned?).

In the study of the Cement, Oklahoma, petroleum and natural gas field, a positive correlation of radon with gas seepage rate and carbon-13 gradient has been noted. A brief cross-section sketch of the area is shown in figure 10.9. Here, the rocks overlying the reservoir are, of course, impermeable, so the radon and other gases bypass this cap rock and migrate toward the surface at the edges of the cap as shown. As various workers have pointed out, while the raw correlations can be used for purposes of drill hole siting, the reasons for the association are not at all well understood at present.

EPITHERMAL METAL DEPOSIT PROSPECTING

Radon anomalies are associated not only with uranium deposits. Many hydrothermal systems from which metal deposits form carry above-background amounts of radon in solution. When the deposits form, residual fluids are thus often enriched in radon. Leakage of this radon to the surface may form subtle but distinct haloes over the metal deposits.

FIGURE 10.9
Use of radon release for oil and gas exploration. See text for detail.

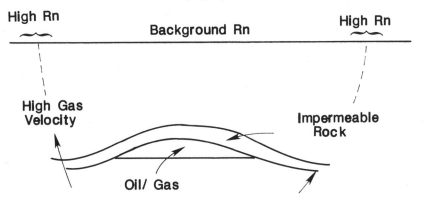

These haloes often result from favorable structural sites for gases, and these same sites are often favorable for metal precipitation. Some workers note that radon can "see" through pediment cover to underlying epithermal gold deposits. Untested, or at least presently unproven, is whether radon haloes may be as effective as conventional geochemical prospecting tools.

CONCLUSIONS

There are many potential uses of monitoring radon emanations other than for indoor concentrations. In this chapter, I have briefly discussed the use of radon emanations for prospecting for uranium, oil, gas, geothermal resources, and epithermal metal ore deposits and for locating faults or fractures in conjunction with siting studies and earthquake predictions. Radon monitoring is also used in sites of abandoned or active uranium mill tailings for health assessment (Morgan et al. 1987) and for many other purposes not covered in this book.

The message is clear: Radon is often a useful tracer for a wide variety of applications, and such uses are increasing rapidly.

11

Radon Versus Other Risks

The focus of this book is on the indoor radon problem yet, I feel it beneficial to include a brief overview on risk from many sources, *not just* that from indoor radon. My approach is not to use sophisticated statistics and modeling, but rather to use the results of other such studies to provide the reader with an empirical, somewhat common sense list of risks. This list is primarily concerned with the United States, but it could just as easily apply to many other nations as well.

THE NATURAL RADIATION BACKGROUND

Radiation is with us all the time. We are radioactive. We ingest appreciable amounts of radioactive ^{40}K (potassium-40) and ^{14}C (carbon-14, or radiocarbon), as well as minute amounts of radioactive uranium, thorium, radium, and rubidium. We inhale radioactive ^{222}Rn and the radon daughters, other naturally occurring radioactive gases and particulates, and small amounts of radioactive materials present in the atmosphere from nuclear weapons use and testing of decades ago. In addition, we are bombarded by cosmic rays, and we receive radiation from the ground, consumer products such as smoke detectors, color television, and other sources. Medical x-rays may be of importance, too. The NRC estimates an average of 70 to 80 mrem per year for average Americans due to x-rays for diagnostic and therapeutic exposures. In table 11.1, I have listed some of the common sources of radiation to which the average citizen in the United States is exposed.

RADON AND THE NATURAL RADIATION BACKGROUND

The role of indoor radon on natural radiation background was largely ignored until the early 1980s. Since then, Dr. Richard Toohey of Argonne National Laboratory and others have been calling attention to the fact that *40 percent* of a typical American's background radiation may come from indoor radon, and is shown in figure 11.1. The remaining 60 percent derives from all the other items tabulated in tables 11.1 and 11.2. Table 11.2 is a personal radiation exposure chart. Note that even at low indoor radon levels of 1 to 2 pCi/L, this amounts to 100 to 200 mrem radiation dose received over a year. For a home with 8 pCi/L, this means an additional 800 mrem/year, on top of all the other items listed in the tables 11.1 and 11.2.

The risk from radon relative to other common sources of risk is given in figure 11.2. Here it is noted that even at about 3 pCi/L, the risk is roughly equal to that from 200 chest x-rays per year. At about 14 pCi/L, the risk is the same as that in someone who smokes a pack of cigarettes per day. At 200 pCi/L, the risk of a fatal lung cancer is 44 to 77 percent—or the same as the risk from smoking more than four packs of cigarettes per day. The message is clear: High indoor radon levels can be very hazardous to your health.

Now, however, is the time to ask if overall high background radiation really is hazardous to one's health? *If* the background radiation is from sources other than indoor radon, a slightly elevated level may not be hazardous at all. We live in a bath of radiation all our lives, and some of this low-level radiation is beneficial in that it kills germs, bacteria, etc. Without such radiation, there would be widespread disease and death.

However, let me call attention to an interesting study carried out in the People's Republic of China (1980), in which the government selected two groups of people from Guangdong Province in Southeast-Central China, and where the overall population is about 1.6 million. The area is agricultural, and the people are not transient. Geologically, there is a highland area underlain by monzonite (Th, U)-bearing granitic rocks next to a valley district. The people living on the highlands receive about 2.5 to 3 times the background radiation (excluding indoor radon) than

FIGURE 11.1

Radon exposure—USA average. From Toohey (1987).

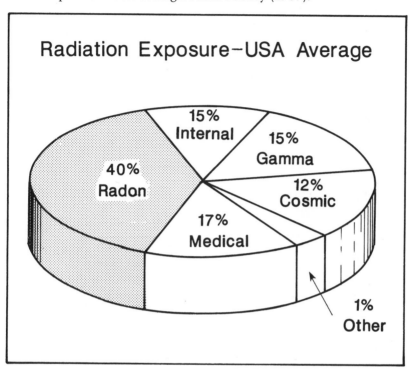

Radiation Exposure–USA Average

40% Radon

15% Internal

15% Gamma

12% Cosmic

17% Medical

1% Other

TABLE 11.1

Natural Radiation Background: United States Inhabitants

Source	Percent of Total
Natural background (sun, air water, food, soil)	67.60
Medical irradiation (diagnostic x-rays and radiation treatment)	30.70
Fallout from nuclear weapons	0.60
Consumer products (watches, smoke detectors, color TV)	0.50
Occupational exposure (x-ray technicians, and welding inspectors)	0.35
Nuclear power plants (all facets of nuclear fuel cycle)	0.10
Coal power plants (all facets of coal fuel cycle)	0.15

Source: Modified from DOE 1986.

TABLE 11.2
Personal Radiation Chart

Activity	mrem/year
Cosmic rays: 28 mrem at sea level. Add 1 mrem for each 100 feet above sea level to 28; i.e., for Denver, CO, at 5200 feet elevation, add 52 to 28 = 80.	28 + __ = __
Food, beverage (average in U.S.)	28
Medical x-rays: add 40 per chest or body x-ray; 14 per dental x-ray (note: in the United States the average is 70–80 mrem/year), add higher figure if you've had radiation treatment.	—
Building materials: for brick, cement, etc., add 100 mrem; for wood, add 5 mrem.	—
Ground radiation (average in U.S.)	26
Nuclear weapons fallout	4
For each 1500 miles flown in a commercial jet: add 1 mrem.	—
For each person you spend 8 hours per day with: add 0.1 × number of days.	0.1 × __ = __
If you live within five miles of a nuclear or coal power plant: add 0.3 mrem.	—
If you live more than five miles from a nuclear or coal power plant: add 0.	—
Indoor radon: *if* your dwelling is close to the mean for the United States (1–2 pCi/L), add 100–200 mrem; if your indoor radon level is higher, add more (i.e., 3 pCi/L – 300 mrem, etc.).	—
Watching color TV: (estimated) add 1–2 mrem/yr	1–2
Luminous watch dial: add 3 mrem/yr.	—

Sources: Compiled from Brookins (unpub.) and other sources.

the valley group (400 and 150 mR/y, respectively). Participants in the two groups were selected on the basis of equal health, ability to trace their ancestry back several generations, and similar lifestyles. Some 70,000 individuals were selected for each group, a staggeringly large figure when one considers the few tens to perhaps hundreds of former miners, occupational workers, etc., studied in most radiation investigations. These 140,000 people wore dosimeters for the five-year period 1974–1979, taking them off only when bathing or at night, when they were placed close by. The dosimeters were then collected, the measurements were made, and the data was studied extensively.

FIGURE 11.2.
Radon risk evaluation chart. From EPA (1984).

Radon Level (pCi/L)	Estimated Fatal Lung Cancers per 1000	Comparable Exposure Levels	Comparable Risk
200	440–770	1000 times average outdoor level	More than 60 times non-smoker risk
			Four Pack/day smoker
100	270–630	100 times average indoor level	20,000 chest x-rays/yr
40	120–380	100 times average outdoor level	Two pack/day smoker
20	60–210		
			One pack/day smoker
10	30–120	10 times average indoor level	
			Five times non-smoker risk
4	13–50		
		10 times average outdoor level	200 chest x-rays/yr
2	7–30		
			Non-smoker risk of fatal lung cancer
1	3–13	Average indoor level	
			20 chest x-rays/yr
0.2	1–3	Average outdoor level	

Source: EPA (1984)

There is only one conclusion of importance. There were *no* statistically significant different health effects, including cancers, between the two groups. If the high background radiation were effective in causing birth defects, leukemia, etc., it should have shown up, but it did not. In fact, the high radiation group was slightly more healthy than the low-level radiation group. Again, this is the most convincing study of its kind ever carried out.

In the United States, there is also no correlation of increases in cancer or other disease with background radiation. For example, the greatest longevity in the United States is in North Dakota, yet this state has high background radiation, and the EPA reports very high radon levels as well.

For indoor radon, however, the picture is different. While the background gamma and beta radiation we receive may not be dangerous to us, except for unwise exposure to the sun (in which case melanoma may result), the long-term effects of alpha-emitting radon daughters may be of concern. Chapter 4 discusses this in detail.

INDOOR RADON AND OTHER SOURCES OF RISK

In this section, I wish to call attention to risk from many different sources. Some of these sources are natural, some are anthropogenic; the latter include both voluntary and involuntary sources. Thus, on the one hand, if one chooses to smoke, one is exposed to risks voluntarily. If, on the other hand, one is passively exposed to smoke, one is an unwilling and involuntary participant to the risk source.

My purpose here is not to try to bury the reader with statistics or pros and cons on choices. All the figures I use below are published in the listed literature, and interested readers should examine the original sources to address their criticisms. I think, however, the figures are of interest, especially when putting indoor radon into perspective with other common sources of risks.

The figures in table 11.3 are taken from several references. Some estimated figures are included therein and should be interpreted accordingly. My approach here is for a rough reconnaissance on the subject only. In addition, I do not include long-term health estimates of potential risk. Instead, I list the actual or estimated fatalities per year for each source of risk. I have included the references for inspection.

Since many Americans are concerned about the health effects of our various energy options, and since nuclear energy—a potential source of radiation—is of prime media concern, a few comments on energy sources are in order.

Table 11.3 lists premature deaths from the sulfur dioxide emitted by fossil fuels. The possibility of 50,000 deaths may come as a surprise to some readers. Yet this figure, which carries an uncertainty of ± 100 percent, from a Brookhaven National Laboratory study published by the Office of Technology Assessment in 1982, is very alarming. The culprit

TABLE 11.3
Some Sources of Risk in the United States (Fatalities/Year)

Sources (Risk)	Fatalities Per Year	Reference
Smoking (lung cancer)	150,000	Upton 1982
Alcoholic beverages	100,000	Upton 1982
Motor vehicles	50,000	Upton 1982
Fossil fuels emit sulfur dioxide (premature deaths)	50,000[a]	OTA 1982
Indoor radon (lung cancer)	20,000	NAS 1988
Handguns	18,000–30,000	Upton 1982, SAUSA 1987
Chronic bronchitis	15,000	SAUSA 1987
Falls	12,000	SAUSA 1987
Electrocutions	14,000	SAUSA 1987
Benign neoplasms	6,500	SAUSA 1987
Drownings	5,300	SAUSA 1987
Conflagrations	5,200	SAUSA 1987
Passive smoking (lung cancer)	5,000	Repace and Lowrey 1985

[a] Although this figure of 50,000 fatalities has been published by OTA 1982, it has an error of perhaps ±80 percent. See text for details.
Source: Office of Technical Assessment 1982; Repace and Lowrey 1985; SAUSA 1987; Upton 1982.

for many of these 50,000 fatalities per year (with a large error—but remember large errors can be off in *both* directions) is coal. The petroleum and natural gas industries get some credit, but coal is the dominant contributor here. The large figure of 15,000 chronic bronchitis fatalities per year also includes those cases of coal miners suffering from anthracosilicosis (black lung), massive pulmonary fibrosis, coal miner's pneumonoconiosis, and others. Coal also shares some blame for contributing CO_2 to the atmosphere and exacerbating the **GREENHOUSE EFFECT**, acid mine drainage, acid rain, and lots of other things not only deleterious to the environment but also with pronounced adverse health effects. The petroleum and natural gas industries also contribute their share.

So, what about nuclear energy? Coal is used as the energy source in part of the nuclear fuel cycle, so I estimate about 80 fatalities per year from all nuclear sources due to the coal used—well below the fatalities per year from coal (50,000–60,000), from petroleum (2,900), and from gas (900).

The handgun fatalities per year vary widely from year to year, but the figure is always high.

Falls and accidental electrocutions cause a number of deaths that in many instances could be avoided. So, too, most drownings and deaths by fires could be circumvented.

It is interesting to note that the leading killers listed in table 11.3 are those that involve a choice such as smoking and drinking alcoholic beverages. With smoking, the figure is closer to 380,000 if cardiovascular disease is added to lung cancer. The alcoholic beverage figure includes, of course, those who are killed by drunk drivers, and the motor vehicle figure includes some of the alcoholic beverage figure. These three sources are known killers on a large scale, and not a whole lot of progress has been made in trying to reduce these deaths. Lowering the speed limit to 55 mph did save lives, but since it was unpopular the interstate speed limit has again been raised to 65 mph in many areas. Therefore, the number of traffic deaths has risen accordingly.

Smoking also affects nonsmokers. It has been calculated that there are 5,000 fatal lung cancers per year due to passive smoke. Further, since there is a multiplicative effect between indoor radon and smoke, either will enhance the other's adverse health effect.

The major killer from natural causes, though, is the indoor radon source. Due to the effect of alpha particles in the lungs, the incidence of fatal lung cancers from this source is high. No other natural source is documented as a killer in such proportions at the present time.

CONCLUSIONS

Estimates of background radiation in the United States and elsewhere in the world prior to the early 1980s are systematically low by a few to several hundred millirem. This results from the failure of such background radiation studies to include radon in the estimates made. It is now estimated that some 40 percent of one's background radiation is due to indoor radon, and this figure may be higher in areas of high radon emanation to the atmosphere.

Several sources of risk have been presented in this short chapter. It is significant that indoor radon lung cancer fatalities fall in the top five categories of risk to public health.

12

Some Conclusions and Recommendations

In the preceding chapters, I have examined several facets of the indoor radon problem and other related topics. The coverage has not been intended to be in-depth or highly rigorous, rather I have sought to present an overview of the problem. In this short chapter, I will focus on some obvious conclusions, and then finish with recommendations for actions responsible federal and state agencies as well as the public might take.

CONCLUSIONS

The following conclusions are reached based on evaluation of the available literature on radon and related topics and from other comparable studies.

1. Uranium and thorium are widely distributed in the rocks and soils of the earth's crust. Thus, parent materials for daughter radon isotopes are also available worldwide.
2. The most important radon isotope from a health viewpoint is ^{222}Rn. Its decay products, especially ^{218}Po and ^{214}Po, can have a pronounced adverse effect on lung tissues, leading to lung cancer in many cases.

3. Radon isotopes from the ^{232}Th and ^{235}U decay chains, ^{220}Rn (thoron) and ^{219}Rn (actinon), are less important than ^{222}Rn, although in some thoriferous areas ^{220}Rn may be of concern.

4. Radium isotopes are easily fixed in some soils. Thus, for ^{226}Ra, parent to ^{222}Rn, this fixation stabilizes the source.

5. Radon release from soils and rocks is complex. From rocks, uranium fixed in minerals along grain boundaries and in defects is susceptible to leaching, thus releasing radon trapped therein. In soils, radon release is affected by moisture content, permeability, porosity, temperature, and other factors. Each site may have distinct characteristic rock and soil properties and must be evaluated accordingly.

6. Radon entry into dwellings usually occurs through cracks, joints, pipe fittings in walls, loose sealants or caulking around windows, and so on.

7. Radon may also enter from building materials, especially rock and rock materials. Hollow blocks are especially prone to radon release. Cements in which coal-produced fly ash is used may have especially high uranium content and thus potentially high radon release rates.

8. Some well waters, especially in high uranium terrains, are high in radon. This radon is released when the waters are heated.

9. Some radon in dwellings is from outside air, but the amounts are usually low when compared to soil-derived radon.

10. There are many radon measuring devices. Those in which a time-integrated average is taken give the best data. Sampling carried out over periods of one week or less should be viewed with caution since radon levels may vary considerably over short periods of time.

11. Consumers should pay special attention to lists of reputable firms with radon measuring services. There may exist a large number of unreputable firms in the radon measurement business.

12. Radon mitigation can range from simple to complex. Passive or forced air ventilation may work to reduce radon levels in many instances; other more complex and more expensive methods of ventilation, i.e., sub-slab suction and wall-in-pipe or baseboard drains, may be necessary to reduce radon levels to acceptable limits. Sub-slab suction methods appear to work extremely well.

13. Consumers should pay close attention to contractors who offer radon mitigation services and take pains to ensure that the firms are

listed with Better Business Bureaus, and the EPA and are properly bonded. Their familiarity with the radon problem is a must.

14. It is not possible to estimate reliably the number of single family dwellings with indoor radon levels above 4 pCi/L air in the United States. Estimates range from seven to 30 percent (see chapter 7).

15. Throughout the world it may be possible to define areas of potentially high radon release based on geologic, geochemical and airborne survey studies. Geologic favorability maps (see chapter 2) for uranium as well as airborne radiometrics are available, as are uranium sediment and water data for much of the United States. These should be utilized.

16. In nations other than the United States, many thousands of radon measurements are available, especially from Canada, Sweden, and some other European countries. For much of the world, however, indoor radon data are lacking.

17. In the arid to semi-arid parts of the United States, there is a disproportionately low number of indoor radon measurements. In the southwestern United States, this is especially true, and preliminary work (see chapter 7) suggests a high number of homes with high indoor radon due to high uranium, radon in soils, ease of radon release, excess insulation (often with solar system units) in homes, and possibly building materials. Possibly 30 percent of homes are rated above 4 pCi/L.

18. Studies of radon health effects in uranium and other metal miners argue convincingly for a dose-adverse health effect relationship between radon and lung cancer. These studies led the National Academy of Science and others to estimate lung cancer fatalities per year due to indoor radon at about 20,000, thus making indoor radon the second leading cause of fatal lung cancer per year in the United States.

19. There is probably a multiplicative effect between smoking and indoor radon, but the data are not sufficient to estimate the combined effect.

20. Small cell carcinoma types of lung cancer may be more prevalent from indoor (and other) radon than from other causes, but this remains to be unequivocally demonstrated (see chapter 4).

21. Relative to other risks (see chapter 11), indoor radon may be the largest single natural cause of fatalities per year in the United States.

22. Use of radon emanations for a wide variety of economic prospecting and other studies, including search for deposits of uranium, epithermal metals, oil and gas, and geothermal resources, is widespread. Radon emanations are also useful for studying sites for environmentally sensitive facilities, earthquake prediction, and related purposes.

23. Despite good intentions, federal and state agencies in the United States have been only partially successful in advising and educating the public about indoor radon.

24. To more fully assess and evaluate the impact of indoor radon on humans, more large animal health studies are needed.

25. There is strong multiplicative correlation between smoking and indoor radon for smokers as well as between passive smoking and indoor radon for nonsmokers. Therefore, reduction of indoor radon levels will be helpful and healthy for everyone.

RECOMMENDATIONS

1. *Education.* More widespread education is needed at all levels. Not only the indoor radon problem, but all aspects of radiation—from background radiation to nuclear energy and even nuclear weapons —need to be examined. There should be federal, state, and private funds made available for such studies. Due in part to poor communication from scientists to the public, but due more to sensationalized and often inaccurate materials presented by the media as fact, there are widespread misconceptions about all facets of nuclear and radiation issues—including indoor radon.

2. *Consumer awareness.* I have observed and studied indoor radon detection practices in various parts of the country, and I am led to recommend strongly that consumer groups become more involved. Inexpensive radon detectors, such as charcoal canisters and alpha-track detectors, are sold for prices many times over their cost. In some cases, radon measurements may not be made at all. I recommend that all communities have a list of reputable firms or individuals providing services for radon detection and mitigation in order to protect the public from unscrupulous individuals and groups that may be promoting unreliable (if not false) radon detection and mitigation.

3. *Research on health effects of radon.* Despite voluminous studies on the health effects of indoor radon, there remain large uncertainties in estimating health effects from this source. Long-range laboratory studies on large animals could provide some of these badly needed data, unpopular as this type of work may be. More follow-up studies on miners and those in other occupations exposed to above-normal radon amounts are also advocated.

4. *Basic research on radon budget and releases from rocks and soils.* Research in these areas is elegant but minimal. Increases in funding, especially to qualified university and other groups outside the national laboratories in the United States, could greatly improve our knowledge of radon. We need more data for radon as a function of rocks and soils and their permeabilities, porosities, moisture contents, etc., although some research is already being done on a somewhat limited scale. In addition, we need more careful soil and rock characterization, including microradiaugraphy and detailed mineralogy and geochemistry (including, where warranted, scanning electron microscopy). Only by rigorously quantifying soil and rock will we be able to develop reliable means for zeroing in on areas of potential radon problems.

5. *Use of available maps, surveys, etc., for radon prediction.* Despite the huge volume of data and maps compiled by the government as part of NURE project, very little attempt has been made to integrate these studies with those covering the indoor radon problem. Already existing maps, data sets, and airborne radiometric surveys should be made available so that areas of potentially high indoor radon can be defined. The EPA, NRC, and DOE should recognize the availability and usefulness of these sources and act accordingly.

6. *Radon mitigation.* For the most part, radon mitigation techniques that work quite well have been developed. However, in the cases of high soil radon emanations into dwellings, mitigation techniques are of limited use. The development of techniques to remove radon (and decay products) already in a dwelling as well as sealing future construction should be given greater priority by the research community. Further, more cost-effective modifications of traditional radon mitigation techniques should be undertaken.

7. *Basic research on radon in building materials and water.* Despite much research in these areas, there remains uncertainty concerning

the role that building materials and waters may play in indoor radon. This is very true in areas where possibly high-uranium-bearing materials are incorporated into or are used directly for building materials. Waters are generally not a problem, but private wells still make up a significant part of domestic water sources in the United States. Relatively few have been studied.

8. *Statistically significant sampling across the United States for radon budget.* I recommend that a national program of radon sampling be initiated and that it emphasize geologic, hydrologic, radiometric, population, and other criteria so as to best obtain a realistic sample for the United States. This can and should be done with advisement and funding from the EPA, HEW, and NIH.

9. *Awareness of Federal agencies (EPA, etc.) and others on the science underlying risk assessment of representative topics.* It is strange that while the documented risk from asbestos is slight, it is funded handsomely. The media as well as the EPA still think in terms of thousands of deaths per year from asbestos instead of the few hundred that are actually documented. Yet radon, with 20,000 fatalities, receives only a fraction of funding. Not only is it important to educate the public about radon and other risks, but it is evident that within the EPA, there is room for tremendous improvement in the education of their spokespersons and regulators.

FINAL STATEMENT

The indoor radon problem is very real. Indoor radon and its decay products account for some 20,000 lung cancer fatalities per year in the United States alone. Worldwide, the figure is much higher. Unlike environmental problems where anthropogenic causes can be identified and regulated, indoor radon comes from the earth. Only by an appreciation of how and where radon is distributed—and can be released—can people come to understand this pressing problem. The intent of this book has been to introduce the reader to the many facets of this problem and to help in the overall goal of radon education.

Appendix: Addresses of State Agencies Responsible for Radon Information

The following appendix is taken from the EPA ("Radon Reduction Techniques for Detached Houses: Technical Guidance", EPA/625/5-86/019, 1986). It is presented here for the convenience of the readers of this book. The names of the key individuals for each State have been omitted in the event of changes in personnel. The last few entries in this appendix are for EPA regional offices. In the event that information cannot be obtained from the State address, the reader is encouraged to try the regional EPA numbers and addresses.

STATE RADIOLOGICAL HEALTH PROGRAM OFFICE CONTACTS (RAD85)

Homeowners and contractors should first contact their State official, listed below, if they require assistance in interpreting the material in this manual or for further support in resolving indoor radon problems.

Alabama

Director
Division of Radiological Health
State Department of Public Health
State Office Building
Montgomery, Alabama 36130
Business: 205/261-5315

Alaska

Chief
Radiological Health Program
Department of Health & Social Service
Pouch H-06F
Juneau, Alaska 99811-9976
Business: 907/465-3019

Arizona

Director
Arizona Radiation Regulatory Agency
925 South 52nd Street, Suite 2
Tempe, Arizona 85281
Business: 602/255-4845

Arkansas

Director
Division of Radiation Control &
Emergency Management
Department of Health
4815 West Markham Street
Little Rock, Arkansas 72201
Business: 501/661-2301

California

Chief
Radiological Health Branch
State Department of Health Services
714 P Street, Office Bldg. 8
Sacramento, California 95814
Business: 916/322-2073

Colorado

Director
Radiation Control Division
Department of Health
4210 East 11th Avenue
Denver, Colorado 80220
Business: 303/320-8333, Ext. 6246

Connecticut

Director
Radiation Control Unit
Department of Environmental
Protection
State Office Building
165 Capital Avenue
Hartford, Connecticut 06106
Business: 203/566-5668

Delaware

Program Administrator
Office of Radiation Control
Division of Public Health
Department of Health & Social
Services
Cooper Building, Cooper Square
Post Office Box 637
Dover, Delaware 19901
Business: 302/736-4731

District of Columbia

Administrator
Department of Consumer &
Regulatory Affairs
Service Facility Regulation
Administration
614 H Street, N.W., Room 1014
Washington, D.C. 20004
Business: 202/727-7190

Florida
Director
Office of Radiation Control
Department of Health &
Rehabilitative Services
1317 Winewood Boulevard
Tallahassee, Florida 32301
Business: 904/497-1004

Georgia
Director
Radiological Health Section
Department of Human Resources
878 Peachtree Street, Room 600
Atlanta, Georgia 30309
Business: 404/894-5795

Hawaii
Chief
Noise and Radiation Branch
Environmental Protection and Health
Services Division
Department of Health
591 Ala Moana Boulevard
Honolulu, Hawaii 96813
Business: 808/547-4383

Idaho
Program Manager
Radiation Control Section
Department of Health and Welfare
Statehouse Mail
Boise, Idaho 83720
Business: 208/334-4107

Illinois
Manager
Office of Environmental Safety
Department of Nuclear Safety
1035 Outer Park Drive
Springfield, Illinois 62704
Business: 217/546-8100
800/672-3380 (Toll Free In State)

Indiana
Chief
Radiological Health Section
State Board of Health
1330 West Michigan Street
Post Office Box 1964
Indianapolis, Indiana 46206
Business: 317/633-0152

Iowa
Director
Environmental Health Section
Iowa Department of Health
Lucas State Office Building
Des Moines, Iowa 50319
Business: 515/281-4928

Kansas
Manager
Bureau of Air Quality and Radiation
Control
Department of Health and
Environment
Forbes Field, Building 740
Topeka, Kansas 66620
Business: 913/862-9360

Kentucky
Manager
Radiation Control Branch
Cabinet for Human Resources
275 East Main Street
Frankfort, Kentucky 40621
Business: 502/564-3700

Louisiana
Administrator
Nuclear Energy Division
Office of Air Quality and Nuclear
Energy
Department of Environmental
Quality
Post Office Box 14690
Baton Rouge, Louisiana 70898-4690
Business: 504/925-4518

Maine

Assistant Director
Division of Health Engineering
157 Capitol Street
Augusta, Maine 04333
Mailing Address: State House,
Station 10
Augusta, Maine 04333
Business: 207/289-3826

Maryland

Administrator
Community Health Management
Program
Department of Health and Mental
Hygiene
O'Conor Office Building
201 West Preston Street
Baltimore, Maryland 21201
Business: 301/225-6031

Massachusetts

Director
Radiation Control Program
Department of Public Health
150 Tremont Street, 7th Floor
Boston, Massachusetts 02111
Business: 617/727-6214

Michigan

Chief
Division of Radiological Health
Bureau of Environmental and
Occupational Health
Department of Public Health
3500 North Logan Street
Post Office Box 30035
Lansing, Michigan 48909
Business: 517/373-1578

Minnesota

Chief
Section of Radiation Control
Environmental Health Division
Minnesota Department of Health
717 Delaware Street, S.E.
Post Office Box 9441
Minneapolis, Minnesota 55440
Business: 612/623-5323

Mississippi

Director
Division of Radiological Health
State Department of Health
3150 Lawson Street
Post Office Box 1700
Jackson, Mississippi 39215-1700
Business: 601/354-6657

Missouri

Chief
Bureau of Radiological Health
1730 East Elm Plaza
Post Office Box 570
Jefferson City, Missouri 65102
Business: 314/751-8208

Montana

Chief
Occupational Health Bureau
Department of Health and
Environmental Sciences
Cogswell Building
Helena, Montana 59620
Business: 406/444-3671

Nebraska

Borchert, Harold R., Director
Division of Radiological Health
Department of Health
301 Centennial Mall, S.
Post Office Box 95007
Lincoln, Nebraska 68509
Business: 402/471-2168

Nevada
Supervisor
Radiological Health Section, Health Division
Department of Human Resources
505 East King Street
Carson City, Nevada 89710
Business: 702/885-5394
800/992-0900 (Toll Free In State)

New Hampshire
Program Manager
Radiological Health Program
Post Office Box 148
Concord, New Hampshire 03301
Business: 603/271-4588

New Jersey
Chief
Bureau of Radiation Protection
Division of Environmental Quality
Department of Environmental Protection
380 Scotch Road
Trenton, New Jersey 08628
Business: 609/530-4000
800/648-0394 (Toll Free In State)

New Mexico
Acting Chief
Radiation Protection Bureau
Environmental Improvement Division
Department of Health and Environment
Post Office Box 968
Santa Fe, New Mexico 87504-0968
Business: 505/827-2948

New York
Director
Bureau of Environmental Radiation Protection
State Health Department
Empire State Plaza, Corning Tower
Albany, New York 12237
Business: 518/473-3618

North Carolina
Chief
Radiation Protection Section
Division of Facility Services
Department of Human Resources
Post Office Box 12200
Raleigh, North Carolina 27605-2200
Business: 919/733-4283

North Dakota
Director
Division of Environmental Engineering
Department of Health
Missouri Office Building
1200 Missouri Avenue
Bismarck, North Dakota 58501
Business: 701/224-2348

Ohio
Director
Radiological Health Program
Department of Health
246 North High Street
Post Office Box 118
Columbus, Ohio 43216
Business: 614/466-1380
800/523-4439 (Toll Free In State)

Oklahoma
Chief
Radiation & Special Hazards Service
State Department of Health
Post Office Box 53551
Oklahoma City, Oklahoma 73152
Business: 405/271-5221

Oregon

Manager
Radiation Control Section
State Health Division
Department of Human Resources
1400 Southwest Fifth Avenue
Portland, Oregon 97201
Mailing Address:
State Health Division
Post Office Box 231
Portland, Oregon 97207
Business: 503/229-5797

Pennsylvania

Director
Bureau of Radiation Protection
Department of Environmental
Resources
Fulton Building, 16th Floor
Third and Locust Street
Harrisburg, Pennsylvania 17120
Mailing Address:
Post Office Box 2063
Harrisburg, Pennsylvania 17120
Business: 717/787-2480
800/237-2366 (Toll Free In State)

Puerto Rico

Director
Radiological Health Division
G.P.O. Call Box 70184
Rio Piedras, Puerto Rico 00936
Business: 809/767-3563

Rhode Island

Chief
Division of Occupational Health
and Radiation Control
Department of Health
Cannon Building, Davis Street
Providence, Rhode Island 02908
Business: 401/277-2438

South Carolina

Chief
Bureau of Radiological Health
South Carolina Department of Health
and Environmental Control
2600 Bull Street
Columbia, South Carolina 29201
Business: 803/758/8354

South Dakota

Radiation Safety Specialist
Licensure and Certification Program
State Department of Health
Joe Foss Office Building
523 East Capital
Pierre, South Dakota 57501
Business: 605/773-3364

Tennessee

Director
Division of Radiological Health
TERRA Building
150 9th Avenue, N.
Nashville, Tennessee 37203
Business: 615/741-7812

Texas

Chief
Bureau of Radiation Control
Department of Health
1100 West 49th Street
Austin, Texas 78756-3189
Business: 512/835-7000

Utah

Director
Bureau of Radiation Control
State Department of Health
State Office Building, Box 45500
Salt Lake City, Utah 84145
Business: 801/538-6734

Vermont
Director
Division of Occupational and
Radiological Health
Department of Health
Administration Building
10 Baldwin Street
Montpelier, Vermont 05602
Business: 802/828-2886

Virginia
Director
Bureau of Radiological Health
Division of Health Hazard Control
Department of Health
109 Governor Street
Richmond, Virginia 23219
Business: 804/786-5932

Washington
Chief
Office of Radiation Protection
Department of Social & Health
Services
Mail Stop LE-13
Olympia, Washington 98504
Business: 206/753-3468

West Virginia
Director
Industrial Hygiene Division
151 11th Avenue
South Charleston, West Virginia
25303
Business: 304/348-3526

Wisconsin
Chief
Radiation Protection Section
Division of Health
Department of Health & Social
Services
Post Office Box 309
Madison, Wisconsin 53701
Business: 608/273-5181

Wyoming
Chief
Radiological Health Services
Division of Health & Medical Services
Hathaway Building
Cheyenne, Wyoming 82002-0710
Business: 307/777-7956

U.S. ENVIRONMENTAL PROTECTION AGENCY PROGRAM RESPONSIBILITIES

Guimond, Richard J., Director
Criteria and Standards Division
Office of Radiation Programs (ANR-460)
Environmental Protrection Agency
401 M Street, S.W.
Washington, D.C. 20460
FTS: 557-9710
Commercial: 703/557-9710

Health effects, measure-
ment protocols, contractor
proficiency program action
level guidance, quality as-
surance

Bliss, Wayne A., Director

ORP Las Vegas Facility
Environmental Protection Agency
Post Office Box 18416
Las Vegas, Nevada 89114
FTS: 545-2476
Commercial: 702/798-2476

Sampling and analysis field evaluation

Porter, Charles R., Director
Eastern Environmental Radiation Facility
Environmental Protection Agency
1890 Federal Way
Montgomery, Alabama 36109
FTS: 534-7615
Commercial: 205/272-3402

Sampling and analysis field evaluation

Craig, A. B., Deputy Director
Air & Energy Engineering Research
Laboratory (MD-60)
Environmental Protection Agency
Research Triangle Park, North Carolina 27711
FTS: 629-2821
Commercial: 919/541-2821

Radon mitigation research program (new and existing houses)

Cotruvo, Joseph A., Director
Criteria and Standards Division
Office of Drinking Water (WH5500)
Environmental Protection Agency
401 M Street, S.W.
Washington, D.C. 20460
FTS: 382-7575
Commercial: 202/382-7575

Radon and radiation in drinking water

Keene, Bryon E., Chief
Radiation and Noise Branch
Environmental Protection Agency, Region 1
John F. Kennedy Federal Building
Boston, Massachusetts 02203
FTS: 223-4845
Commercial: 617/223-4845

EPA Regional Representative for Connecticut, Maine, Massachusetts, New Hampshire, Rhode Island, and Vermont

Giardina, Paul A.
Environmental Protection Agency
Region 2 (2AIR:RAD)
26 Federal Plaza
New York, New York 10278
FTS: 264-4418
Commercial: 212/264-4418

EPA Regional Representative for New Jersey, New York, Puerto Rico, and the Virgin Islands

Belanger, William
Environmental Protection Agency
Region 3 (3AH14)
6th and Walnut Streets
Philadelphia, Pennsylvania 19106
FTS: 597-9800
Commercial: 215/597-9800

EPA Regional Representative for Delaware, District of Columbia, Maryland, Pennsylvania, Virginia, and West Virginia

Payne H. Richard
Environmental Assessment Branch
Environmental Protection Agency,
Region 4
345 Courtland Street, N.E.
Atlanta, Georgia 30365
FTS: 257-3776
Commercial: 404/347-3776

EPA Regional Representative for Alabama, Florida, Georgia, Kentucky, Mississippi, North Carolina, South Carolina, and Tennessee

Tedeschi, Pete
Environmental Protection Agency
Region 5 (5AHWM)
230 South Dearborn Street
Chicago, Illinois 60604
FTS: 353-2654
Commercial: 312/353-2654

EPA Regional Representative for Illinois, Indiana, Michigan, Minnesota, Ohio, and Wisconsin

May, Henry D.
Environmental Protection Agency
Region 6 (6T-AS)
1200 Elm Street, Suite 2800
Dallas, Texas 75270
FTS: 729-5319
Commercial: 214/767-5319

EPA Regional Representative for Arkansas, Louisiana, New Mexico, Oklahoma, and Texas

Brinck, William L.
Environmental Protection Agency, Region 7
726 Minnesota Avenue
Kanas City, Missouri 66101
FTS: 757-2893
Commercial: 913/236–2893

EPA Regional Representaive for Iowa, Kansas, Missouri, and Nebraska

Lammering, Milt
Environmental Protection Agency
Region 8 (8AH-NR)
1860 Lincoln Street
Denver, Colorado 80295
FTS: 564-1710
Commercial: 303/293-1700

EPA Regional Representative for Colorado, Montana, North Dakota, South Dakota, Utah, and Wyoming

Duncan, David L.
Environmental Protection Agency
Region 9 (A-3)
215 Fremont Street
San Francisco, California 94105
FTS: 454-8378
Commercial: 415/974-8378

EPA Regional Representative for American Samoa, Arizona, California, Guam, Hawaii, and Nevada.

Cowan, J. Edward
Environmental Protection Agency
Region 10 (Mail Stop 532)
1200 Sixth Avenue
Seattle, Washington 98101
FTS: 399-7660
Commercial: 206/442-7660

EPA Regional Representaive for Alaska, Idaho, Oregon, and Washington

Glossary

Actinon: An isotope of radon, ^{219}Rn.

AEC: Atomic Energy Commission.

Air exchange ratio: The rate at which fresh air coming into a home completely replaces the home's existing air supply.

Air filtration: A radon reduction technique in which air is passed through high efficiency filters or electronic devices, which collect dust, etc., and to which radon is affixed.

Air-to-air heat exchanger: A ventilation device used to reduce radon levels in homes by retaining indoor heat while exchanging indoor air and accompanying pollutants for fresh outdoor air.

Alpha particle: Two neutrons and two protons bound as a single particle that is emitted from the nucleas of certain radioactive isotopes during their decay; a helium nucleus ($_2^4$He^{2+}).

Alkali elements: Sodium, potassium, rubidium, cesium, lithium.

Alkaline earth elements: Beryllium, magnesium, calcium, strontium, barium, radium.

Anion: (see ion).

Atom: Fundamental particle of all matter, consisting of one or more protons and zero or more neutrons in a nucleus surrounded by a number of electrons equal to the number of protons.

Apatite: Mineral; $Ca_5(Po_4)_3(OH,F)$.

A-horizon: Uppermost zone of soil; zone of leaching.

Background radiation: Radiation from radioactive substances; includes cosmic rays, soil-rock-water radiation, air radiation, etc.

Barite: Mineral; $BaSO_4$.

Basement pressurization: A radon reduction technique whereby the air pressure in the basement is increased to raise it above the soil radon air pressure from the soil.

Basic rocks: Rocks with low silica (SiO_2) content, from about 45 to 55 percent SiO_2.

Bauxite: Residual laterite rich in aluminum.

Becquerel: Unit of radioactivity; 1 becquerel (Bq) = 1 disintigration per second (dps).

Bentonite: Mineral; complex Mg-rich hydroxylated alumino silicate clay mineral.

Beta particle: Particle given off during certain radioactive decay; an electron.

BEIR: Biological effects of ionizing radiation.

B-horizon: Soil zone of accumulation.

Calcite: Mineral; $CaCO_3$.

Caliche: Hard, indurated, calcite-rich layers in soil in arid terrain.

Cation: (see ion)

Charcoal canister: A radon measurement device that absorbs dust and accompanying radon.

Coffinite: Mineral; $USiO_4$.

Curie: Unit of radioactivity = 3.7 × 10^{10} disintigrations per second (dps).

C-horizon: Zone of active attack; soil developing on bedrock.

Daughter product: Isotope formed as a result of radioactive decay.

Diffusion: Random path followed by very small particles due to impact of surrounding molecules.

Dose equivalent: The quantity that, for radiation, expresses the assumed effectiveness of dose on a common scale for all kinds of ionizing radiation. Expressed in rems (1 rem = 100 ergs/gram approximately) or sieverts (1 Sv = 100 rem).

DOE: U.S. Department of Energy.

DNA: Deoxyribonucleic acid; basic building block molecules in animals.

Drain tile ventilation: Radon reduction technique whereby radon is drawn away from a home by an underground drainage system surrounding the home.

Electron volt: A unit of energy = 1.6×10^{12} ergs = 1.6×10^{19} joules.

Emanation coefficient: Factor for reporting emanation of radon into the air.

EPA: U.S. Environmental Protection Agency.

Epithelium: Membranous cellular tissue that covers the surface of some organ or a part of the body.

Epigenetic: In geology, refers to something formed after the host rock has formed.

Epithermal: In geology, refers to shallow, fairly low temperature ore deposits.

ERDA: U.S. Energy Research and Development Administration.

Evaporites: Deposits of salt formed by evaporation of a closed basin of water.

Evapotranspiration: Process by which leaves and other vegetative matter lose water to the atmosphere.

Gamma ray: Short wavelength electromagnetic radiation of nuclear origin from some radioactive decay process. Close to an x-ray.

General sealing: Radon reduction techniques that involve barriers between radon source and a home's living space.

Greenhouse effect: Build-up of CO^2 and other gases to form thermal blanket around the earth such that reflected solar radiation is trapped; causes increase in temperature of the earth. Due primarily to the burning of fossil fuels.

Gypsum: Mineral; $CaSO_4.2H_2O$.

Half-life: The time required for 50 percent of a radioactive parent isotope to decay to a daughter isotope.

Hematite: Mineral; Fe_2O_3.

HEW: U.S. Department of Health, Education and Welfare.

Home ventilation: Radon reduction technique where a home's existing air supply is replaced by fresh, outdoor air; may be done passively (opening windows) or by forced methods (use of fans).

Hormesis: Applies to a threshold dose below which risk is very low, or reverses (i.e., very high or very low dose cause adverse health effects).

Ion: An atom that has gained or lost electrons. A cation is an atom that has lost one or more electrons, hence has a positive charge; an anion is an atom that has gained one or more electrons, hence has a negative charge.

Ionic radius: The radius, in angstroms (10–8 cm), of an ion.

Isotopes: Different atoms of the same element with different numbers of neutrons in the nucleus.

ICRP: International Commission on Radiation Protection.

KGRA: Known geothermal resource area.

Latent period: Period of time between exposure and development of disease; i.e., after exposure to high radon dose, there is a latent period of several years before lung cancer developes.

Lifetime risk: The lifetime probability of dying of a specific disease.

Linear dose model (linear hypothesis): Model that postulates that the excess risk is lineraly proportional to dose.

Lithophile: In geology, those elements that have an affinity for silicon and oxygen and from mainly silicate minerals.

Metallogenic Province: Area or region of the earth enriched in one or more metals well above background.

Mill tailings: Waste rock and chemicals from metal milling processes; usually disposed near site of milling.

Monazite: Mineral, complex rare earth element, Th phosphate.

Mrem: One one-thousandth rem.

Multiplicative model: Model in which independent risk factors interact so that the combined risk is the product of the relative risks of each factor alone; i.e., smoking and indoor radon for lung cancer.

NAS: National Academy of Science.

NCRP: National Council on Radiation Protection.

Neoplasms: Any new and/or abnormal growth, such as a tumor.

NIH: U.S. National Institute of Health.

NRC: U.S. Nuclear Regulatory Commission.

NURE: U.S. National Uranium Resource Evaluation.

Pegmatite: Igneous rock; last product of some magmatic crystallization. Pegmatites commonly contain very large crystals in equilibrium with very small crystals and possess unusual chemistries and mineralogies. Often rich in uranium.

Permeability: Rate of fluid transmittal in rocks.

Phosphate slag: Waste materials from phosphate mining that are sometimes used as foundation material in construction work. Often contains uranium.

Picocurie (pCi): One trillionth of a curie.

Porosity: Void space in rocks.

ppb: Parts per billion; i.e., 1 ppb = 0.0000001 weight percent.

ppm: Parts per million; i.e., 1 ppm = 0.0001 weight percent.

Progeny: Decay products resulting after a series of radioactive decay; i.e., the radon daughters are the radon progeny.

Quadratic-dose model: Model that assumes that excess risk is proportional to the square of the dose.

Rad: A unit of absorbed dose; commonly assumed equal to the rem.

Rem: A unit of absorbed dose; roughly equal to 100 ergs/gram.

Redox front: Boundary or zone between oxidizing and reducing conditions in rocks.

Roll front: In geology, that roll-like shape of uranium-mineralized rocks at or near the redox front due to formation of uranium minerals.

Soil ventilation: Radon reduction technique whereby radon is ventilated or drawn away from a home before it can enter.

Squamous cell carcinoma: Cancer composed of cells that are plate-like or scaly.

Stochastic: Describes random events that lead to effects whose probabilities of occurrence in an exposed population is a direct function of dose. No threshold is assumed.

Sub-slab ventilation: Radon reduction technique whereby pipes are placed in stone aggregates beneath a home's basal floor to ventilate radon.

Thorium: Element.

Thoron: Radon isotope; ^{220}Rn, formed from (ultimately ^{232}Th.

Threshold: For health purposes, that level of concentration below which there is no meaureable or actual health effect.

Ultrabasic rocks: Rocks with less than 45 percent SiO_2.

Unattached fraction: Fraction of radon daughters (usually ^{218}Po) that has not yet attached to a particle. As a free ion, it has a high probability of being held within the lung and depositing alpha energy when it decays.

Uraninite: Mineral; UO_2.

Uranium: Heaviest naturally occurring element.

USGS: U.S. Geological Survey

Vadose zone: Zone in rocks or soil between the water table and the surface.

Wacke: An impure, clay-rich sandstone.

Working level (WL): A unit of measure for exposure to radon decay products where one WL is approximately equal to 200 pCi/L.

Working level month (WLM): A unit of measure for measuring cumulative exposure to radon; 1 WLM equals exposure to 1 WL for 173 hours (i.e., hours worked for 1 month on the average).

Zircon: Mineral; $ZrSiO_4$.

References

Introduction

NAS, 1988. Health effects of radon and other internally deposited alpha-emitters: National Academy Press, Washington, D.C., 602 pp.

Nazaroff, W.W. and Nero, A.V., eds. 1987. Radon and its decay products in indoor air: Wiley-Interscience, New York, 518 pp.

1. Radon Units, Standards, and Related Items

Alter, H.W., and Oswald, R.A. 1987. Nationwide distribution of indoor radon concentrations: A preliminary data base: Journal Air Pollution Control Assoc., vol. 37, p. 227–232.

Brookins, D.G. 1988. The indoor radon problem: Studies in the Albuquerque, New Mexico area: *Environ. Geol. and Water Science,* vol. 9, pp. 315–334.

Cohen, B.L. 1979. Radon: Characteristics, natural occurrence, technological enhancement, and health effects: *Prog. Nuc. Energy,* vol. 4, pp. 1–24.

EPA, 1986. Guidance for dealing with radon: *EPA Journal,* vol. 12, pp. 12–13.

Kirsch, L.S. 1986. Liability for injuries caused by high levels of radon: *in Radon Update,* Aug. 1986. Terradex Corp., p. 1.

NAS, 1988. Health risks of radon and other internally deposited alpha-emitters: BEIR IV: National Academy of Science Press, Washington, D.C., 602 pp.

Nero, A.V. 1986. The indoor radon story: *Tech. Review,* Jan. 1986, pp. 28–40.

Nero, A.V. 1987. Radon and its decay products in indoor air: An overview in Radon and its decay products in indoor air (W.W. Nazaroff and A.V. Nero, eds.), Wiley-Interscience Pubs., New York, pp. 1–56.

2. Uranium and Thorium in the Earth's Crust

Brookins, D.G. 1976. Uranium deposits of the Grants, New Mexico, mineral belt: U.S. Energy Research and Development Administration Rpt. GJBX-16(76), 153 pp.

Brookins, D.G. 1979. Uranium deposits of the Grants, New Mexico, mineral belt (II): U.S. Department of Energy Rpt. 76–0029E, 411 pp.

Brookins, D.G. 1984. Geochemical aspects of radioactive waste disposal: Springer-Verlag Pubs., New York and Heidelberg, 347 pp.

Clark, S.P., Peterman, Z.E. and Heier, K.S. 1966. Abundances of uranium, thorium and potassium: *in* Handbook of physical constants, revised ed. (S.P. Clark, Ed.), Geol. Soc. Amer. Mem. 97, pp. 521–541.

Cohen, B.L. 1979. Radon: characteristics, natural occurrence, technological enhancement, and health effects: *Prog. Nuc. Energy*, vol. 4, pp. 1–24.

Morgan, et al., National Council on Radiation Protection, 1975. Natural background radiation in the United States: National Council on Radiation Protection and Measurements Rpt. 45, Washington, D.C. 163 pp.

Myrick, T.E., Berven, B.A., and Haywood, F.F. 1983. Determination of concentration of selected radionuclides in surface soil in the U.S.: Health Physics, vol. 45, pp. 631–636.

Nazaroff, W.W., Moed, B.A. and Sextro, R.G. 1987. Soil as a source of indoor radon: Generation, migration and entry: in Radon and its decay products in indoor air (W.W.. Nazaroff and A.V. Nero, eds.), pp. 57–112.

Rich, R.A., Holland, H.D. and Petersen, U. 1977. *Hydrothermal uranium deposits:* Elsevier Sci. Pub. Co., Amsterdam: 264 pp.

Rogers, J.J. and Adams, J.A.S. 1967. Uranium: *in Handbook of Geochemistry,* vol. 2 (K.H. Wedepohl, ed.), Springer-Verlag Pubs., Berlin, Chapter 2, 50 pp.

Scott, R.C. and Barker, F.B.,, 1962. Data on uranium and radium in ground waters in the United States: U.S. Geol. Surv. Prof. Paper 426, 115 pp.

Shearer, S.D., and Sill, C.W. 1969. Evalution of atmospheric radon in the vicinity of uranium mill tailings: *Health Physics,* vol. 17, pp. 77–88.

Travis, C.C., Cotter, S.J., Watson, A. P., Randolph, M.L., Mcdowell-Boyer, L.M. and Fields, D.E. 1979. A radiological assessment of radon-222 released from uranium mills and other natural and technologically enhanced sources: U.S. Nuc. Reg. Com. Rpt NUREG/CR-0573, 216 pp.

United Nations, 1977. Sources and effects of ionizing radiation: appendix B: Natural sources of radiation: Rpt. of the United Nations Committee on the Effect of Atomic Radiation, General Assembly, 32 Session, United Nations, New York, 1977.

U.S. Congress, 1982. Management of comingled uranium mill tailings, H.A.S.C. No. 97–55,, 97th Congress, 2nd session, Aug. 17–18, 1982.

Wedepohl, H., Ed. 1967. Handbook of Geochemistry: Springer-Verlag Pub . . . Berlin, Chapter 2 (Uranium), 50 pp.

Whittaker, E.J.W. and Muntus, R. 1970. Ionic radii for use in geochemistry: Ceochim. Cosmochim. Acta, vol. 34, pp. 945–956.

Wollenberg, H.A. 1984. Naturally occurring radioelements and terrestrial gamma ray exposure rates: LBL Rpt. 18714, Berkeley, California, 18 pp.

3. Radioactive Decay of Uranium and Thorium

Brookins, D.G. 1984. Geochemical aspects of Radioactive Waste Disposal: Springer-Verlag Pubs., New York and Heidelberg, 347 pp.

Cohen, B.L. 1979. Radon: Characteristics, natural occurrence, technological enhancement, and health effects: *Prog. Nuc. Energy*, vol. 4, pp. 1–24.

NAS 1988. Health risks of radon and other internally deposited alpha-emitters; BEIR IV: National Academy Press, Washington, D.C., 602 pp.

Nero, A.V. 1987. Radon and its decay products in indoor air: An overview in Radon and its decay products in indoor air (W.W. Nazaroff and A.V. Nero, eds.), Wiley-Interscience, New York, pp. 1–56.

Schery, S.D. 1985. Measurements of airborn 212Pb and ^{220}Rn concentrations at varied indoor locations within the United States: *Health Phys.*, vol. 49, pp. 1061–1065.

Schery, S.D. 1985. Studies of thoron and thoron progeny: Indicators for transport of radioactivity from soil to indoor air: in Indoor radon, Proc. APCA Int. Spec. Conf., Philadelphia; Air Pollution Control Assoc. 16 pp.

Seelmann-Egebert, Pfennig, G., Münzel, H., Klewe-Nebenius, H., 1981. Chart of the nuclides: Kernforschungszentrum Karlsruhe GmbH, Karlsruhe, FRG. 32 pp.

4. Health Effects of Radon and Radon Progeny

Archer, V.E., Radford, E.P. and Axelson, O. 1978. Factors in exposure-response relationship of radon daughter injury: in Proc. Conf. Workshop on Lung Cancer Epid. in Industr. Appl. of Sputum Cytology: Colo. Schl. Mns. Golden, CO, pp. 324–336.

Beir I. [Biological Effects of Ionizing Radiation] 1972. The effects of populations of exposures to low levels of ionizing radiation: National Academy of Science Press, Washington, D.C., 487 pp.

Cohen, B.L. 1979. Radon: characteristics, natural occurrence, technological enhancement, and health effects: *Prog. Nuc. Energy*, vol. 4, pp. 1–24.

Cothern, C.R. 1987. Estimating the healtgh risks of radon in drinking water: *Research and Technology*, April, 1987. pp. 153–158.

Cross, F.T. 1987. Evidence of lung cancer from animal studies: in Radon and its decay products in indoor air (W.W. Nazaroff and A.V. Nero, eds.), Wiley-Interscience, New Yorik, pp. 373–406.

EPA, 1986. Guidance for dealing with radon: *EPA Journal*, vol. 12, pp. 12–13.

Lundin, F.D., Wagoner, J.K. and Archer, V.E. 1971. Radon daughter exposure and respiratory cancer, quantitative and temporal aspects: Joint monograph no. 1, U.S. Public Health Service.

NAS, 1988. Health risks of radon and other internally deposited alpha-emitters: BEIR IV: Nat. Acad. Press, Washington, D.C., 602 pp.
NCRP, 1975. Natural background radiation in the United States: National Council, Rad. Prot. Rpt. 45, Bethesda, MD, 163 pp.
Saccomanno, G., Yale, C., Dixon, W., Auerbach, O. and Huth, G.C. 1986. An epidemiological analysis of the relationship between exposure to radon progeny, smoking and bronchogenic carcinoma in the U-mining population of the Colorado Plateau, 1960–1980: Health Physics, vol. 50, pp. 605–618.
Svec, J., Kunz, E. and Placek, V. 1976. Lung cancer in uranium miners and long termm exposure to radon daughter products: Health Physics, vol. 30, pp. 433–442.
Toohey, R.E. 1987. Radon vs. lung cancer: New study weighs the risks: Logos (Argonne National Lab.), vol. 5, pp. 7–11.
Weiffenbach, C. 1982. Radon, water, and air pollution: Risks and control: Univ. Maine - Orono, Land and Water Resources Center.

5. Radon Entry

Brookins, D.G. 1977. Uranium abundance in some Precambrian and Phanerozoic rocks from New Mexico: Rocky Mtn. Assoc. Geol. Gdbk. 1977. pp. 353–360.
Brookins, D.G. 1988. The indoor radon problem: Studies in the Albuquerque, New Mexico area: Env. Geology and Water Science, vol. 8, pp. 308–324.
Brookins, D.G. 1989. Mineral and energy resources: Occurrence, exploitation, and their environmental impact: Charles E. Merrill, Columbus, OH.
Brookins, D.G., Wethington, J.A. and Merklin, J.F. 1981. Exploration, reserve estimation, mining, milling, conversion, and properties of uranium: NSF Sponsored Learning Module, Kansas State University, Manhattan, KS, 164 pp.
Bruno, R.C. 1983. Sources of indoor radon in houses: A review: Jnl. Air Pollution Control Assoc, vol. 33, pp. 105–109.
Duncan, D.L., Gesell, T.F. and Johnson, R.H. 1976. Radon-222 in potable water: Proc. Health Phnysics Soc. Tenth Mid-Year Sym.: Natural Radioactivity in Man's Environment, conf 761031, Rennselear Polytech. Inst. Press, Troy, NY, 340–355.
Fleischer, R.L. 1987. Moisture and ^{222}Rn emanation: Health Physics, vol. 52, pp. 797–799.
Gesell, T.F. and Prichard, H.M., The technologically enhanced natural radiation environment: Health Physics, vol. 28, pp. 361–366.
Hultqvist, B. 1956. Studies on naturally occurring ionizing radiations: Kungl. Svenska Veten. Handl., 4, serien, Band. 6, Nr. 3, Stockholm (in Swedish).
Kahlos, H. and Asikainen, M. 1980. Internal radiation doses from radioactivity of drinking water in Finland: Health Physics, vol. 39, pp. 108–113.
Michel, J., and Jordana, M.J. 1987. Nationwide distribution of radium-228,

216   REFERENCES

radium-226, radon-222 and uranium in ground water: *in* Radon in ground-water (B. Graves, ed.,), Lewis Pubs., Chelsea, MI, pp. 227–240.

Morgan, et al., National Council on Radiation Protection, 1975. Natural background radiation in the United States: National Council on Radiation Protection and Measurements Rpt. 45, Washington, D.C. 163 pp.

NAS, 1988. Health risks of radon and other internally deposited alpha-emitters: BEIR IV: National Academy of Science Press, Washington, D.C., 602 pp.

Nazaroff, W.W., Moed, B.A. and Sextro, R.G. 1987a, Soil as a source of indoor radon: Generation, migration, and entry: *in* Nazaroff and Nero, eds. (1987) pp. 57–112.

Nazaroff, W.W., Doyle, S.M., Nero, A.V. and Sextro, R.G. 1987b, Radon entry via potable water: in Nazaroff and Nero, eds. (1987), pp. 131–160.

Nero, A.V. 1987. Radon and its decay products in indoor air: An overview in W.W. Nazaroff and A.V. Nero, eds. (1977), pp. 1–56.

Scott, A.G. 1983. Computer modelling of radon movement: in EML indoor radon workshop (A.C. George, W. Lowder, I. Fisenne, E.O. Knutson and L. Hinchcliffe, eds.), Env. Monitoring Lab. Rpt. 416, New York, pp. 82–100.

Scott, A.G. 1987. Preventing radon entry: *in* Nazaroff and Nero (1987), p. 407–455.

Smith, B.M., Grune, W.N., Higgins, F.B. and Terrill, J.G. 1961. Natural radioactivity in ground water supplies in Maine and New Hampshire: Jnl. Amer. Water Works Assoc., vol. 53, pp. 75–88.

Solley, W.B., Chase, E.B. and Mann, W.B. 1982. Estimated use of water in the United States in 1980: U.S. Geol. Surv. Circ. 1001.

Stranden, E. 1987. Building materials as a source of indoor radon: in Nazaroff and Nero, eds. (1977), pp. 113–130.

Wilkening, M.H. 1985. Radon transport in soil and its relation to indoor radioactivity: *Sci. Tot. Env.*, vol. 45, pp. 2199–226.

Wilkening, M.H., Clements, W.E. and Stanley, D. 1972. Radon-222 flux measurements in widely separated regions: in Natural Radiation environment II (J.A.S. Adams, W.M. Lowder and T.F. Gesell, eds.), Conf. 720805, National Tech. Infor. Serv. Springfield, VA, pp. 717–734.

Wollenberg, H.A. 1984. Naturally occurring radioelements and terrestrial gamma-ray exposure rates: An assessment based on recent geochemical data: Lawrence Berkeley Lab. Rpt. LBL-18714.

6. Radon Detection Methods

Cohen, B.L. and Cohen, E.S. 1983. Theory and practice of radon monitoring with charcoal adsorption: *Health Physics,* vol. 45, pp. 501–510.

EPA, 1986. Interim indoor radon and radon decay product measurement protocols: U.S. Env. Protection Agency Rpt.. EPA 520/1–86–04, 50 pp.

George, A.C. 1984. Passive, integrated measurement of indoor radon using activitated charcoal: *Health Physics,* vol. 46, pp. 501–509.

Kotrappa, P., Dempsey, J.C., Kickey, J.R., and Stieff, L.R. 1988. An electret passive environmental ^{222}Rn monitor based on ionization measurements: *Health Physics,* vol. 54, pp. 47–56.

Nazaroff, W.W. 1987. Measurement techniques: in Radon and its decay products in indoor air (W.W. Nazaroff and A.V. Nero, eds.), Wiley-Interscience, New York, pp. 491–504.

7. Radon Studies in the United States

Alter, H.W., and Oswald, R.A. 1988. Nationwide distribution of indoor radon concentrations: A preliminary data base: Journal Air Pollution Control Assoc., vol. 37, pp. 227–232.

Brookins, D.G. 1977. Uranium abundance in some Precambrian and Phanerozoic rocks from New Mexico: Rocky Mtn. Assoc. Geol. Gdbk. 1977. pp. 353–360.

Brookins, D.G. 1984. Geochemical aspects of radioactive waste disposal: Springer-Verlag Pubs., New York and Heidelberg, 347 pp.

Brookins, D.G. 1986. Indoor and soil Rn measurements in the Albuquerque, New Mexico area: *Health Physics,* vol. 51, pp. 5299–533.

Brookins, D.G. 1988. The indoor radon problem: Studies in the Albuquerque, New Mexico area: *Env. Geology and Water Science,* vol. 9, pp. 32–43.

Ethier, W. 1988. *Radon News Digest,* Sept. 1988. pp. 3–9.

Nazaroff, W.W., and Nero, A.V. 1987, Radon and its decay products in indoor air: Wiley-Interscience, N.Y., 518 p.

Nero, A.V. 1987. Radon and its decay products in indoor air: an overview: *in* Nazaroff and Nero, Eds; pp. 1–56.

8. Remedial Actions Methods in the United States

Brookins, D.G. 1986. Indoor and soil Rn measurements in th. Albuquerque, New Mexico area: *Health Physics,* vol. 51, pp. 5299–533.

Brookins, D.G. 1988. The indoor radon problem: Studies inthe Albuquerque, New Mexico area: *Env. Geology and Water Science,* vol. 9, pp. 32–43.

EPA, 1986. Radon reduction methods: A homeowner's guide: U.S. Env. Prot. Agency Rpt. OPA-86–005, 24 pp.

Jonassen, N., and McLaughlin, J.P. 1987. Removal of radon and radon progeny from indoor air: *in* Radon and its decay products in indoor air (W.W. Nazaroff and A.V. Nero, eds.), Wiley-Interscience, New York, pp. 435–458.

Nero, A.V. 1987. Elements of a strategy for control of indoor radon: *in* Radon and its decay products in indoor air (W.W. Nazaroff and A.V. Nero, eds.), Wiley-Interscience, New York, pp. 459–490.

Rudnick, S.N., Hinds, W.C., Maher, E.F., Price, J.M., Fujimoto, K. Gu Fang and First, M.W. 1982. Effect of indoor air circulation systems on radon decay product concentrations: Final Rpt., EPA Contract 68–00106029, Harvard School of Public Health, Boston.

9. Radon Studies in Other Countries

Akerbolm, G. 1986. Investigationand mapping of radon risk areas: Int. Sym. Geol. Mapping Serv. Env. Planning Commm. Geol. Map World (CGMW), Trondjheim, Norway, May 6–9, 1986. 15 pp.

Aoyama, T., Yonehara, H., Kakanoue, M., Kobayashi, S., Iwasaki, T., Mifune, M., Radford, E.P. and Kato, H. 1987. Long-term measurements of radon concentrations in the living environments of Japan: A preliminary report: in Radon and its decay products: Occurrence, properties, and health effects: Amer. Chem. Soc. Sym. Series 331, Amer. Chem. Soc., Washington, D.DC., pp. 124–232.

Brookins, D.G. 1986. Indoor and soil Rn measurements in the Albuquerque, New Mexico area: *Health Physics,* vol. 51, pp. 529–533.

Green, B.M.R., Brown, L., Cliff, K.D., Driscoll, C.M.H., Miles, C.J. and Wrixon, A.S. 1985. Surveys of natural radiation exposure in UK dwellings with passive and active measurement techniques: *Sci. Total Env.,* vol. 45, pp. 459–470.

Keller, G. and Folkerts, K.H. 1984. Radon -222 concentrations and decay-product equilibrium in dwellings and in the open air: *Health Physics,* vol. 47, pp. 385–398.

Kobal, I., Smodis, B. and Skofljanec, M. 1986. Radon 222 air concentrations in the Slovenian Karst caves Yugoslavia: *Health Physics,* vol. 50, pp. 830–834.

Kobal I., Smodis, B., Burger, J. and Skofljanec, M. 1987. Atmospheric ^{222}Rn in tourist caves of slovenia, Yugoslavia: *Health Physics,* vol. 52, pp. 473–479.

Letourneau, E.G. and Wigle, D.T. 1980. Mortality and indoor radon daughter concentrations in thirteen Canadian cities: in Proc. Spec. Mtg. Assess. Radon and Daughter Exp. and Rel. Biol. Effects, Rome, Italy: RD Press, Salt Lake City, UT, p. 239–251.

Letourneau, E.G., McGregor, R.G. and Walker, W.B. 1985. Design and interpretation of large surveys for indoor exposure to radon daughters: *Radiat. Prot. Dosim.,* vol. 7, pp.303–308.

McAulay, I.R. and McLaughlin, J.P. 1985. Indoor radiation levels in Ireland: *Sci. Total Env.,* vol. 45, pp.319–326.

McGregor, R.G., Vasudev, P., Letourneau, E.G., McCullough, R.S., Prantl, F.A. and Taniguchi, H. 1980. Background concentrations of radon and radon daughters in Canadian homes: *Health Physics,* vol. 39, pp.285–300.

Nero, A.V. 1987. Radon and its decay products in indoor air: An overview: in Radon and its decay products in indoor air (W.W. Nazaroff and A.V. Nero, eds.), Wiley-Interscience, New York, pp.1–56.

Poffijin, A., Marijns, R., Vanmaarcke, H. and Uyttenhuve, J. 1985. Results of a preliminary survey on radon in Belgium: *Sci. Total Env.,* vol. 45, pp.335–342.

Put, L.W. and de Meijer, R.J. 1984. Survey of radon concentrations in Dutch dwellings: in Indoor air: radon, passive smoking, particulates an housing epidemiology, vol. 2 (B. Berglund, T. Lindvall, and J. Sundell, eds.), Swedish Council for Buillding Res., Stockholm, pp.49.

Schmeir, H. and Wicke, A. 1985. Results from a survey of indoor radon exposures in the Federal Republic of Germany: *Sci. Total Env.*, vol. 45, pp.307–314.

Sorensen, A., Botter-Jensen, L, Majborn, B. and Nielsen, S.P. 1985. A pilot study of natural radiation in Danish homes: *Sci. Total Env.*, vol. 45, pp.351–358.

Swedjemark, G.A. and Mjönes, L. 1984a, Radon and radon daughter concentrations in Swedish homes: Radiat. Prot. Dosim., vol. 7, pp.341–348.

Swedjemark, G.A. and Mjönes, L. 1984b, Exposure of the Swedish population to radon daughters: *in* Indoor air: Radon, passive smoking, particulates and housing epidemiology (B. Berglund, T. Lindvall and J. Sundell, eds.), vol. 2, Swedish Cncl. Building Res., Stockholm, pp.37–54.

Wolfs, F., Hofstede, H., De Meijer, R.J. and Put, L.W. 1984. Measurements of radon-daughter concentrations in and around dwellings in the northern part of the Netherlands: A search for the influences of building materials, construction and ventilation: *Health Physics,* vol. 47, pp.271–279.

10. Economic Uses of Radon Emanations

Brookins, D.G. 1984. Geochemical aspects of radioactive waste disposal: Springer-Verlag, New York and Heidelberg, 347 p.

Fisher, J.C. 1976. Application of track etch radon prospecting to uranium deposits, Front Range, Colorado, World Metal Mining Tech. Proc., Joint MMIJ-AIME Meeting, vol. 1, Denver, CO, Sept. 1–3, 1976, pp.95–111.

Fleischer, R.L. and Turner, L.G. 1984. Correlations of radon and carbon isotopic measurements with petroleum and natural gas at Cement, Oklahoma: *Geophysics,* vol. 49, pp.810–817.

Gingrich, J.E. 1983. Radon as a geochemical exploration tool: 10th Int. Sym. Geochem. Explor., Helsinki, Finland, Aug. 29–Sept. 2, 1983. 24 p.

King, C.Y. 1980. Episodic radon changes in subsurface soil gas along active faults and possible relation to earthquakes: *Jour. Geophys. Res.,* vol. 85, pp.3065–3078.

Mansker, W.L. 1987. Radionuclide anomalies in geologic exploration: GEORAD Conf., St. Louis, MO, Apr. 21–22, 1987, p .9.

Mogro-Campero, A. and Fleischer, R.L. 1976. Changes in subsurface radon concentration associated with earthquakes: *Earth Plan. Sci. Lttrs.,* vol. 34, 321–327.

Morgan, G.M., Banerjee, S., Brookins, D.G., Cohen, N., Domenico, P.A., Hirschfield, R.C., James, H.L., Kulp, J.L., Neill, R.H., Shoemaker, E.M. and Wiltshire, S. 1987. Scientific basis for risk assessment and management of uranium mill tailings: National Academy Press, Washington, D.C., 246 p.

11. Radon Versus Other Risks

Brookins, D.G., in press, Mineral and energy resources: Occurrences, exploitation and their environmental impact: Charles E. Merrill Pub. Co., Columbus, OH.

Department of Energy, 1986. Understanding radiation: U.S. Dept. Energy Booklet DOE/NE-0074, 17 p.

Environmental Protection Agency, 1986. Guidance for dealing with radon: U.S. Env. Prot. Agency. *EPA Journal,* Aug. 1986. vol. 12, pp.12–13.

NAS 1988. Health risks of radon and other internally deposited alpha-emitters: National Academy Press, Washington, D.C., 602 p.

Office of Technology Assessment, 1982. The regional implications of transported air pollutants: An assessment of acidic deposition and ozone: Office Technology Assessment, Washington, D.C., July 1982. pp.J-1 - J-16.

People's Republic of China, 1980. Health survey in high background radiation areas in China: *Science,* vol. 209, pp.877–880. (People's Republic of China High Background Radiation Research Group.)

Repace, J.L. and Lowrey, A.H. 1985. A quantitative estimate of nonsmokers' lung cancer risk from passive smoking: *Env. Internat.,* vol. 11, pp.3–22.

SAUSA, 1987. Statistical abstract of the United States 1986. 106th Ed.: U.S. Dept. Commerce, Bur. Census, 984 p.

Toohey, R.E. 1987. Radon vs. lung cancer: New study weights the risks: *Logos,* vol. 5 (Argonne Nat. Lab.), pp.7–11.

Upton, A.C. 1982. The biological effects of low-level ionizing radiation: *Sci. Amer.,* vol. 246, pp.41–49.

Index

A horizon, 86
Accessory minerals, 28
Accidental electrocutions, 190
AEC see United States agencies
Air exchange rate, 150
Air supply, 152
Albuquerque, see New Mexico
Alcoholic beverages, 190
Alpha, 55
Alpha particles, 55, 68, 91, 111; radiation, 56
Alpha track detectors, 93, 107-110, 114, 116, 118, 120, 126, 146, 176, 179; measurements, 116
Alum shale, 6, 30, 106, 165, 167
American Society of Heating, Refrigerating, and Air-Conditioning Engineers (ASHRAE), 17
Anthracosilicosis (black lung disease), 189
Approved devices, 119
Argonne National Laboratory, 137
Arizona, 128, 142
Asbestos, 64
Australia, 35

B horizon, 86
Background radiation, 6, 9, 16, 77, 79, 143, 183, 184, 187, 188, 190, 194; tables, 80, 81, 185
Basalt, 27
Basement versus non-basement radon (table), 123
Basements, 7, 93, 120, 126, 132, 152
Basic research on radon budget and releases from rocks and soils, 195
Baskin, Roberta, 7
Battelle Pacific Northwest Laboratories, 139
Bauxites, 32
Becquerel, 13
Belgium, 169
Bentonites, 32
BETA, 55
Beta radiation, 56, 57, 288
Birth defects and low-level radiation group, 187
Bonneville Power Administration, 143
Bronchitis, 82
Building materials, 133, 192

Evapotranspiration, 39, 40
Exposed earth, covers, 153

Leaching, 27, 29, 192
Leukemia, 187
Los Alamos National Laboratory, 45, 134, 137
Lung cancer, 4, 14, 18-21, 57, 63-79, 137-141, 165, 184, 189-193
Lung cancer in the United States, 64, 84
Lung diseases, 16, 70, 82

Madera Formation, 129
Magma, 27
Maine, 122, 142
Marine phosphate minerals, 36
Marine phosphorites, 31
Maryland, 8, 49, 84, 124, 126, 141, 160
Massachusetts, 142
Measuring devices 107, 137, 147, 192
Measuring Techniques, 1, 109, 118, 137,142
Metallogenic Provinces, 37
Metamorphic Rocks, 32, 105, 142
Miners, 5, 21, 70-77, 82, 143, 173, 174, 186; uranium miners, 4, 63, 73, 139
Minnesota, 142
Missouri, 142
Monitoring for radon, 4-7, 84, 101, 129, 137, 141, 156, 162, 182, 195; see also alpha track
Montana, 1, 36
Mose, Dr. Douglas, 50, 84, 126-128, 159

Nagasaki, 72
Namibia, 35
National Academy of Science, (NAS), 18, 64, 73, 95, 193
National Academy of Science-National Research Council, 15

National Airborne Radiometric Reconnaissance Program (NARR), 49
National Association of Home Builders, 144, 146
National Association of Realtors, 146
National Cancer Institute, 141
National Council on Radiation Protection (NCRP), 16, 70, 78
National Institute of Health, 49
National Programs, 44
National Uranium Resource Evaluation, (NURE), 23, 44, 49, 50, 129, 135, 136, 196
Natural background radiation, see, background radiation
Natural radiation background (table), 185
Netherlands, The, 168
Nevada, 36, 44, 49, 128
New Hampshire, 48
New Jersey, 7, 122, 141, 147
New Mexico, 42, 44, 50, 99, 128, 135, 142, 147, 162; Albuquerque, 16, 17, 44, 83, 93, 101, 142, 149
New York, 7, 41, 102, 122, 139, 141, 147
Newfoundland, 5
Nonradon Lung Diseases of Uranium and Other Metal Miners, 82
North Dakota, 142, 188
Norway, 17, 167
Notation, 9
Nuclear energy and nuclear power plants, 6, 8, 188, 189, 194; weapons, 78, 183
NURE, see, National Uranium Resource Evaluation

Oak Ridge National Laboratory, 45
Ohio, 44
Oregon, 49, 122, 141
Oxidation, 24, 34, 38, 50, 89, 195
Oxyhydroxides, 86

Passive or forced air ventilation, 162, 192
Passive smoke, 21, 64, 188, 190, 194
Pennsylvania, 7, 82, 120, 122, 139, 140-143, 147, 162
Pennsylvania Department of Environmental Resources, 7, 141
Peoples Republic of China, 184
Permeable, 43
Permian, 129
Personal radiation chart, 186
Petroleum, 11, 180, 181, 189
Phosphates 31, 32, 36, 37, 40, 48, 99, 101, 140, 141, 143
Picocuries, 14
Pleistocene, 36
Pneumonia, 82
Porphyry deposits, 35
Precambrian deposits, 35
Prospecting for geothermal resource areas, 178
Prospecting for petroleum and natural gas, 180
Prospecting for uranium deposits, 173
Public health, 7, 15, 54, 84, 105, 124, 144, 190
Pulmonary fibrosis, 82, 189

Radiation, 6-9, 16, 18, 72, 75-81, 114, 114; see also background radiation and radon daughters
Radiation doses to individuals (table), 79
Radioactive decay of uranium and thorium, 53
Radioactive particles and rays, 55
Radioactivity, 13, 14, 100, 169
Radiometric age determinations, 37
Radium as parent to radon, 3, 40, 84; concentration in rocks (table), 90; concentrations in some rocks, 44, 45; content of some common rocks and building material ingredients

(table), 103 health effects, 83; in soils, 41, 49, 89, 134; see also radon
Radium budget, 44
Radium isotopes, 192
Radon, 40, 50, 58, 70-72, 78, 87, 88, 151, 179, 180, 192; in the atmosphere, 41; in building materials, 105, 195; emissions of radon gas above ground, 180 entry: sources and mechanisms, 85 (table), 95, 192, 194; from bedrock, 105; from water, 95, 96, 195; health effects, 193; in lung tissue 4; positive correlation of, 181 research/studies, 127-136, 140, 165; in soils in the United States 42 (table), 46; testing, 144 units and standards, 13
Radon and radon daughters, 2, 14, 38, 53, 58, 67, 70, 107, 110; removal from air, 160
Radon and radon isotopes, 2, 53, 58, 192, half-life, 3
Radon budget, 196
Radon data base summary by state (table), 120
Radon detection methods and testing, 19, 107, 144, 192
Radon emanation coefficients for common building materials (table), 104
Radon measurements and concentrations, 120
Radon mitigation, 192, 195
Radon prediction, 195
Radon progeny integrated sampling units for WL determinations, 114
Radon reduction methods: A homemaker's guide, 140
Radon studies in Albuquerque, New Mexico, 128
Radon studies in the United States, 119
Radon versus other risks, 183

Three Mile Island, 6
Tsivoglou procedure, 113

United Kingdom, 5, 168
United States, 5, 17, 39, 42, 64, 84, 119, 149, 165, 193
United States Agencies Atomic Energy Agency (AEC), 36, 44, 136; Department of Energy (DOE), 45; Energy and Resource Development Administration (ERDA), 44; Geological Survey (USGS), 36, 49
Uranium and thorium in the earth's crust, 23
Uranium concentration in igneous rocks (ppm) table, 28
Uranium decay, 2, 37, 54, 55
Uranium deposits, 33, 45, 170, 173, 176
Uranium in waters over specific rock types (table), 33
Uranium mill tailings, 44, 82; miners 73; mining, 70
Utah, 36, 128

Vein deposits, 35
Ventilation by heat-recovery methods, 152
Ventilation of block walls, 157
Ventilation to the outside, 149
Veterans Administration, 143
Virginia, 8, 35, 49, 124, 141

Wacke, 30
Washington (state), 7, 49, 122, 141
Washington, D.C., 7
Watras, Stanley, 7
Well waters, 192
WJLA-TV, 8
Working Level (Wl), 14; working level month (WLM), 15
Working Level sampling method, 113
World Health Organization (WHO), 17
Worldwide—some typical outdoor radon values (table), 41
Wyoming, 34, 36

Yugoslavia, 1, 169

AcB 9599

TD
885-5
R33
B76
1990